"If anything more sane, more lucid and more illuminating .. has appeared in recent years, I haven't seen it!"
—Professor Eduard C. Lindeman

Here at last is a simple, lucid report, written by an expert economist, on how science is learning to measure and control the economic forces which rule our lives. Cutting through abstract theory, it cogently and concisely discusses the basic facts about production, employment and national income which are essential to understanding the modern world.

Here you will learn exactly what is meant by the national income and what it tells us about the history and structure of American economy; what we know about inflation and depression; the relation between national economies and the world economy; the potentialities of that extraordinary new tool of thought, the Nation's Economic Budget.

These are some of the important subjects discussed in this timely book. In straightforward, simple terms, George Soule urges us to exercise the knowledge we have about our economy to beneficially control it, and to junk the economic myths, assumptions and delusions which have so long produced only economic frustration.

This is a must book for everyone who wants to be well informed about modern life. *"Anyone who knows just a little about economics can learn a lot more by reading this book, and he will be learning about the world we live in, instead of a world projected from an ivory tower."*—E. C. Noyes, *The Nation*

Other MENTOR Books of Special Interest

INTRODUCTION TO

ECONOMIC SCIENCE

BY

GEORGE SOULE

A MENTOR BOOK

Published by THE NEW AMERICAN LIBRARY

*Published as a MENTOR BOOK
By Arrangement with The Viking Press, Inc.*

FIRST PRINTING, JANUARY, 1951
SECOND PRINTING, JULY, 1952
THIRD PRINTING, MAY, 1953
FOURTH PRINTING, JULY, 1954
FIFTH PRINTING, MARCH, 1958
SIXTH PRINTING, AUGUST, 1959
SEVENTH PRINTING, OCTOBER, 1961

MENTOR TRADEMARK REG. U.S. PAT. OFF. AND FOREIGN COUNTRIES
REGISTERED TRADEMARK—MARCA REGISTRADA
HECHO EN CHICAGO, U.S.A.

*MENTOR BOOKS are published by
The New American Library of World Literature, Inc.
501 Madison Avenue, New York 22, New York*

PRINTED IN THE UNITED STATES OF AMERICA

CONTENTS

CONTENTS (Cont.)

1

Economic Fundamentalism

WHILE SCIENCE has tremendously increased man's power over other things, his ability to direct his own conduct has not kept pace with his command of physical forces. Human society has difficulty in choosing appropriate ends and still greater difficulty in achieving aims which are generally regarded as desirable. So wide has become this disparity between man's control over non-human nature and his control over himself that many doubt whether the race can escape self-destruction.

Responsibility for this unhappy state of affairs is often placed upon science itself. Modern man, some believe, has devoted too much time to learning about cold and impersonal facts by processes of logic and too little to the spiritual and moral elements of human life. Those who adopt this view are inclined to regard the pursuit of science as a blind alley, in so far as human happiness is involved. Useful though it may be in adding to material comforts and conveniences or to the arts of war, its achievements are now as embarrassing as would be a Diesel locomotive or a machine gun in the hands of a child in a nursery. The assumption of those affected by the reaction against science is that what mankind needs now most of all is not more knowledge or more skill, but more virtue.

This attitude is nowhere more prominent than in certain popular books on economics. The authors of these books seem to believe that true economics is not an experimental science in the modern sense of the word, but a set of principles and maxims discovered once and for all in former centuries. Like Sunday School teachers in a slum, they bid us return to the classical gospel, so that the balance of prices and production which is supposed to emerge from unrestricted operation of the laws of supply and demand under free enterprise may prevail. The world is all but lost; not a nation in it, with the possible exception of the United States, clings to the faith, and even here false prophets have led us astray. But to these authors, the truth is clear and whole. They seem to believe that we have wandered from the truth only because we did not understand. Patiently they explain it to us all over again, often in words of biblical simplicity. Once more they convict us of sin in a score of public policies, all the way from the protective tariff to compensatory spending by government.

The major admonitions of such books are often as unassailable as the Ten Commandments. In judging the wisdom of any course of action, for instance, we should think not only of its effect on those whom it is supposed to benefit, but of its effect on all others as well. And we should think not only of its temporary result, but of its ultimate consequences. Applying these principles, one has no difficulty in proving that destruction does not add to economic welfare but robs us of wealth; that the ultimate good of workers, as well as of everyone else, depends upon improvement of productivity, even though technological advance may throw some out of work; that we cannot gain in the long run by obstructing imports or by giving away exports; that simply adding to the money supply does not make us rich; that inflation is an abomination; and that high taxes to support governmental activities may contribute less to the public welfare than would the same money spent in private enterprises.

These and other axioms of the same sort are universally accepted by economists as well as by sensible folk untrained in the subtleties of theory. To the economic preacher, it seems that the policies he opposes could be advocated only by persons ignorant of these principles or persons who ignored them. Indeed, few economists of any school would support a general policy of limitation of output, or make-work rules by unions, or opposition to mechanical improvement, or inflation, or the abandonment of saving, or governmental squandering of money. The unwary should not conclude that those who deplore the economic revivalist's position are not respectful of his main principles or disagree with his conclusions at every point. But they do disagree that letting things alone will produce the desired results.

What actually is the matter with the school of reasoning which is thus content to return to orthodox first principles? Why has it seemed necessary to revise not the aims, but the application, of the classical concepts?

The Fallacy of Concealed Premises

The first fallacy may be called the fallacy of concealed premises. In the classical economic writers, it was masked by an oft-repeated phrase—*ceteris paribus,* other things being equal. For the sake of clarity, attention was concentrated on a single chain of cause and effect, while the writer assumed that nothing else in the field of forces changed or differed from his assumptions. Sometimes these assumptions were stated, often they were merely implied. But in the real world "other things" do not remain the same, and often differ so far from the assumptions of the writer that if they are taken into account, they upset his conclusion. No good scientist thinks any more of cause and effect as a single chain; science now thinks in terms of a field of forces and of events in that field.

An example first invented by Bastiat and repeated in

a recent book by Henry Hazlitt[1] is an excellent illustration of this fallacy. The breaking of a show window is welcomed by the crowd in the street because the replacement of the window will supply employment. The crowd forgets that the fifty dollars which the merchant must spend for the new window otherwise would have been spent for a new suit, which also would have furnished employment. The community has lost one window or its equivalent, and nothing can compensate for that fact. Labor will be diverted from making suits to making windows. This parable is then applied to explode the belief that war, despite its destructiveness, brings economic benefits.

Now, as a matter of fact, the recent war *did* cause increased employment, as everybody, including Hazlitt, would have to admit if he had his eye on the facts rather than on the theory. Not only that, but in spite of the immense destruction which sopped up so much labor and materials, the goods consumed by the population were, as a whole, larger in quantity than in the years before the war, when unemployment was prevalent.

The trouble with the story about the broken window lies in its unstated and assumed premises. It assumes that the merchant would spend his fifty dollars for a suit or something else if he did not have to buy the window. It assumes that he could not buy the suit and the window too, either by taking the extra money from under the mattress or borrowing it from a bank. It assumes, furthermore, that everybody is already employed, so that labor could not make the new suit if the window had to be made. In the case of the war, these assumptions were all contrary to the facts. The United States did not spend all it could until it had to spend for destruction. The extra money required could be borrowed. There were millions of unemployed, whose potential labor was being

[1] *Economics in One Lesson* by Henry Hazlitt (New York: Harper and Brothers, 1946).

wasted until they were put to work as a result of war demand.

The lesson is not that we should favor war for economic reasons, though when we look at the whole picture, instead of just part of it, it is easier to see why people on occasion believe destruction will benefit them. The lesson is that we should learn how to use our productive capacity without war. Hazlitt is content merely with the axiom that destruction is destruction.

The Fallacy of Missing Quantities

It is characteristic of the popular preachers of classical economics that they often don't take the pains to see what bearing statistical facts have upon their conclusions. Of course, statistics can be misinterpreted, and some statistics are inaccurate representations of the facts they are supposed to report. Statistics should be handled with conscience and care. Nevertheless, the truth emerges, not from contempt for facts, but from respect for them as a means of discovering the actual behavior of the economic world. Your devotee of traditional theory, on the contrary, begins by deducing his principles from a few simple premises. Then he "interprets" the statistics in the light of the "basic principles" he has "learned." Now, in any respectable science, a theory deduced from postulates is regarded not as the final truth, but as a hypothesis to be tested by experience or observation. Its logic may be perfect, granted its assumptions, but still it may not be an accurate representation of nature. Good theory is not independent of the behavior of nature. All scientific laws are, to be sure, simplified abstractions, and variations are almost always discovered in applying them. But careful study of such variations is always in order, because they may indicate flaws in the hypothesis, or may involve such large aberrations as to show that the "law" is irrelevant to the real world. Selecting statistics to prove a hypothesis, or interpreting them on the basis of learned principles, is not the method of scientific discovery. Science has more respect for facts than that.

The example of the broken window is an instance of the error which may result from inattention to statistics. An illustration of their misuse to prove a preconceived dogma has to do with railway wages in the United States. It is a basic principle of classical economics that high prices restrict demand. A frequent inference from this principle is that high wages (wages being regarded as the price of labor) diminish the demand for workers and so reduce employment. To support this conclusion, it is pointed out that in recent years the number of railway workers has gone down while their wages per hour have risen. But nothing is said about the great increase in output per man-hour on the railways, another statistical fact of first importance. This held in check unit labor costs while hourly wages rose; it has made possible the handling of an increasing amount of railway transportation with fewer men. One cannot account for decreased railroad employment without considering this and a dozen other factors.

The Fallacy of Omitted Factors

This may be regarded as a special case of the concealed premises, but it is so important and so characteristic of the school of economic fundamentalism that it is worth separate mention. In applying principles to concrete cases—not merely in formulating them, mind you—the writers of this school ignore the kind of economic world we live in. They assume without further ado that perfect competition exists, that prices and production respond readily to demand, that the more efficient competitors will survive, that labor and capital are highly mobile. This sort of assumption is the basis of their general conclusion that intervention by government or unions is, as a rule, both unnecessary and harmful and that socialism is bound to be less efficient than private enterprise. Occasionally they admit exceptions, but these exceptions are not taken seriously in arriving at their recommendations.

Though your popularizer of classical "law" disapproves restriction of output and price-rigging in general terms, he seldom takes seriously a major development of American economic history—the growth of concentration of business control, its effect on irregularity of production and on price rigidity, its power as a pressure group, its influence on the volume of investment, on the growth and distribution of the national income. Maybe our economy acts just as if competition were as prevalent, and shifts of capital and labor as easy, as Adam Smith thought they ought to be. Or, to be fair, maybe our economy would act in this way if only government, organizations of labor, and other pressure groups would stop interfering. But it is hazardous to assume so without even mentioning monopoly and monopolistic competition. This is only one example of the total disregard of many important factors, such as periodic depression and chronic unemployment.

The Fallacy of Separate Pieces

Modern science does not proceed merely by analyzing separate little pieces of its subject and then by pulling out one of the pieces to apply to a particular problem of practice, forgetting the rest. It not only strives to put the pieces together, as in a picture puzzle, but it also sees that the whole picture has an organic meaning which is more than a collection of its parts—a "Gestalt."

Nobody could be more wholehearted than the opponent of governmental deficits in describing the evils of inflation. Yet he denounces price-control and rationing after a war. He seems to detect no inconsistency in these two positions. The opposition to price-control is based on the argument that a free interplay of prices and profits will lead to the most efficient satisfaction of demand. Yet the description of inflation shows clearly that this does not happen when inflationary forces are at work. Why does it not occur to him that price-fixing and rationing may be employed to moderate inflation? He just doesn't put them together.

Your economic fundamentalist is firm about the desirability of increasing production, and indicates that he knows it has increased in the past and may increase in the future. Yet any enlargement of the volume of purchasing power at any time, whether by governmental borrowing or credit expansion, seems to him "inflationary." When he discusses such measures he assumes that the extra spending will never increase production, but only boost prices. But during a slump we may have idle capacity and unemployed labor. At such a time, might not a larger volume of spending employ them without raising prices? This possibility he never considers; he does not bring the two ideas together. He grossly misrepresents the compensatory-spending school by implying that what they want is rising prices. Government spending, he assumes, will always divert production from private channels, never increase its total.

If he paid more attention to recorded experience, he might see the necessity of putting his pieces together.

The Fallacy About Human Behavior

Underlying every other defect of the economic preacher's view is his misunderstanding of human motivation. It never seems to occur to him to ask why, in spite of the fact that Adam Smith attempted in 1776 to instruct people how they should organize human activities, and on the whole did a better and more celebrated job than has since been done, the economic behavior of the world is, by his standards, more iniquitous than ever.

Perhaps motives are different from those assumed in the classical analysis. For some reason the experience of people does not lead them to accept the approved recommendations. Can we rely on an individualistic society to pursue self-interest by competition and ingenuity and hard work, and to stop pursuing it by combinations, restrictions, demands for hand-outs and special favors? Hundreds of thousands of farmers will not resignedly accept bankruptcy when deflation hits them. When a 1932 arrives, fifteen million unemployed workers will not passively wait for the slow working out of automatic equilibrium by the kind of economic regime which produced their distress.

We may cheerfully admit that no matter what prompts human beings to act as they do, informed and intelligent persons ought not to support remedies which will make things worse instead of better. But is it not the responsibility of the scientist who aspires to influence conduct to find out how and why the real world behaves as it does, instead of sitting off and offering preachments to it which are seldom accepted? We do not lecture the steel which fails to withstand strain and stress when bridges tumble down. In such a case we dig not merely into our mathematics but into the surrounding forces and the properties of the materials with which we have to work.

Not Less Science, But More

The remedy for man's failure to control his own behavior lies, not in abandoning science for dogma, but in more and better science. This is no disparagement of virtue. Goodness is needed, and so are codes of ethics. But neither can help much without realistic understanding. If expositors of economics want to preach, they should be sure that they are able to recommend a feasi-

ble and beneficial mode of conduct. The kind of economics preached by the economic revivalists is not based
upon a mature enough science to be of much use.

During the past twenty or thirty years economics as
a science has taken a long leap ahead. It has been enabled to do so, not because economists in general have
become more clever, or because those few economic
geniuses who have appeared tower over their predecessors
as does a Newton or an Einstein. It has done so mainly
because for the first time in history economists have been
provided with large-scale and reasonably accurate access
to the facts of economic behavior. They have begun to
learn from the world of nature, like any normal scientist,
instead of merely from a few scattered observations and
imagined instances as in the more remote past.

During World War I an enormous amount of statistical
material, hitherto inaccessible, began to be gathered in
connection with the process of governmental control
which was found necessary. In subesquent years, compilation of such facts was continued and expanded, both by
government and by private organizations. Research agencies, bent upon a scientific approach, strove to put this
material in order and interpret it. The depression of the
1930's, followed by World War II, accelerated the
process.

For a time, it looked as if the meaning of the facts was
being forgotten or overwhelmed in the mere bulk of the
statistics turned up. The mine was operating full speed,
but the huge piles of ore were not being smelted and
turned into the desired instruments of thought and control. The scientific economic workers were regarded by
their more elegant and gentlemanly colleagues who dealt
in traditional abstractions as grimy and dim-visioned
miners, grubbing in dark shafts and piling up mountains
of ill-assorted facts without contributing to any broad
view. But the work went on, and presently meaning

began to be introduced. At last there are hints of ways in which the statistical ore may be put to use.

The process is a long way from being completed yet, but it has gone far enough so that at least it ought to be reported. Much is known that was unsuspected by those who wrote when economics was more a philosophy than a science. The new knowledge, and the possibilities of using it, cannot be explained in quite such neat and elementary syllogisms as the parables of primitive economics. Nevertheless, like any body of scientific knowledge, it is worth some effort to master. Lay readers have waded through hundreds of technical pages about discoveries of physics—like relativity and atomic energy— which are far more difficult, and less capable of use by the citizens in the process of conducting their businesses or casting their votes. The purpose of this book is to explain something of the new science of political economy.

2

Keeping Books for All the People

OF ALL THE RECENT ADVANCES in economic knowledge, none is capable of being put to more uses than the studies of the national income. Before these figures were available, it was impossible to tell with any accuracy whether the people of a nation were becoming worse or better off in a material sense, or, if there was progress, what the rate of progress was. Obtaining that information is far from exhausting the uses to which national income figures can be put. They show also what is happening to the various industries, to farmers, wage-earners, property owners, professional people. They are indispensable in making federal budgets and levying taxes. If the tax authorities know roughly how large an income we have and how it is distributed, they can much better calculate the yield of any income tax. Finally, the figures are capable of being put to use in planning for the future, as we shall see in later chapters.

Before it is possible to understand how the national income figures are used, it is necessary to know what they are and something about how they are obtained and put together.

Figures for the national income, past, present and future, appear frequently in the newspapers. So commonly accepted is the term that most people do not realize that it is comparatively new, and that they do not know exactly what it means. Before 1919 there were no good figures for the size of the national income in the United States or for its change from year to year. Nobody knew in detail how it was composed or how it was distributed. Even the word itself was not carefully defined. During the period between the wars, laborious work by statisticians and economists gradually developed the idea and gave it substance.

Much of the pioneer work was done by the staff of a non-governmental research agency, the National Bureau of Economic Research. Later, the United States Department of Commerce took over the job, and in 1947 revised its tabulations. There are gaps in the basic data; estimates have to be made, and the final totals include a margin of error. Yet the several investigators in the field arrive at substantially similar results, and for many practical purposes the figures are accurate enough. Certainly they are far better than unsupported guesses. They are often particularly useful in following changes over a period of time.

Anybody who wants to know the basic facts about an economic enterprise consults a report of its operations prepared by accountants. National income accounting applies similar methods to the affairs of all the people in a nation. The total national income is usually stated in figures of so many billions that they lose meaning to persons accustomed to count their incomes in hundreds or thousands. But these enormous totals are far from the whole story. The businessman who inspects the financial report of a corporation does not stop when he looks at the figures of gross or net income. He also inspects the columns carefully to see how these figures were derived. Likewise, the totals of national income are what emerge after detailed calculations which in themselves reveal some of the most significant facts about the economy.

It has been said that study of the national income is

the branch of economics which corresponds to anatomy in the study of the human body. Anatomy reveals the structure of the organism. It shows the size, composition, and interrelationships of its various parts. Anyone who knows anatomy can go on to study how these parts operate and how they develop. The study of living processes is called physiology. Classical economics attempted to describe the physiology of economic processes before the facts of economic anatomy were known. It talked about the relationship among prices, demand and supply, the functions of saving and capital, the operation of markets and similar subjects, but it talked about them in a world of abstractions. That is one reason why its conclusions often seem so remote and difficult to apply.

Of course, we must have an economic physiology if we are to understand and control our institutions, but we are likely to develop a much better physiology now that the major facts of economic anatomy have become known and roughly measurable.

What the National Income Is

What exactly is meant by the national income? In a sense this term is a figure of speech, since the nation as a whole does not have an income which could be put down on an income-tax return, as is the income of an individual or a corporation. The national income is not, for instance, what a nation earns above its expenses by buying and selling abroad. It is estimated by a process of social accounting which regards the population of a nation as a unit. It is thought of as the sum of all incomes.

Now let us see how the reckoning of the national income looks for a given year, let us say 1939, the last year of peace before the outbreak of World War II. At the beginning, for simplicity, it is better to list only the major totals, as derived from the most readily available figures.

Income Received (1939)	*(Millions of Dollars)*
First we have the compensation of employees	47,820
Next there is the net income of independent business or professional men, such as farmers, doctors, small businessmen or partners	11,282
Many people receive income from rent, and this must be listed	3,465
And we must not forget the profits of corporations (part of this is distributed and part is not, but both parts must be counted)	5,753
Finally, there is the net interest received by individuals	4,212
Since nobody receives income from any other source for current production, the total is the national income	72,532[1]

Income Equals Product

Simon Kuznets, one of the greatest American authorities on the subject, defines the national income as "the net product of, or net return on, the economic activity of individuals, business firms, and the social and political institutions that make up a nation."

There lurks in this definition an extremely important equation. The national income is the net product of economic activity. At the same time, it is the net return from economic activity. In other words, the amount paid for everything that is produced for sale is equal to the sum of the incomes of all the people. This equation, as will be explained later, leads to important conclusions.

[1] There are certain statistical refinements employed in deriving these figures from the raw data which may be of interest.

The compensation of employees includes not merely what they receive directly in wages and salaries, but indirect payments by their employers, such as contributions for social insurance or pension funds and compensation for injuries.

Profits of business as here listed exclude change in the value of inventories arising from shifts in prices. Such changes do not really constitute income, in the sense of something produced.

The figures given for business profits are for the profits *before* payment of taxes on that income. While the owners of the business do

It is easier, at the beginning, to see why output equals income if we think of income in physical terms. We spend our dollars for loaves of bread, houses, automobiles—these things constitute our real income. We can buy what the nation produces and no more, no matter how many dollars we may collectively have. In the case of services, such as those performed by lawyers, teachers or doctors, the service output is obviously the same as the service income of the recipients. The barber, let us say, produces one haircut; at that moment the customer receives one haircut.

Suppose we do not spend some of our dollars for either goods or services but save them. Might not our total money income, in that case, be larger than the total value of the nation's product? The answer is no, not at any given moment. Where did we get the money? It came from somewhere along the course of production, and it was added into the value of the product. Of course, goods may remain unsold; their prices may have to be reduced. But if that happens, it changes the money incomes of the producers at the same time.

Income has to be stated in dollars rather than in goods

---

not retain that part of the profit which is taken by the government in taxes, it is part of the national income.

The figure for rental income which people receive includes not merely payments from tenants, but an estimate of the net rental value of houses occupied by their owners. Use of the houses is a substantial part of the owners' real income, and is so regarded in our government's social accounting.

The item for interest does *not* include interest received on government bonds. While people actually receive this money, it does not represent current production for the most part, since the debt was incurred mainly for war. To include this interest would make the post-war income seem larger than it is, in terms of actual product.

Finally, there are certain payments received, called transfer payments, which are not included in the national income. These are payments for which no services are currently rendered, and consequently do not represent production—payments such as old-age pensions, unemployment benefits and gifts.

if we are to deal in totals at all, since houses, loaves of bread, and automobiles cannot be added together to produce a total which means anything. Attach their prices to them, and it is possible to get the sum.

Another definition of the national income, supplied by the Department of Commerce, calls it "the aggregate earnings of labor and property which arise from the current production of goods and services by the nation's economy." This definition immediately goes on to recognize the fact that income equals product by adding, "Thus, it measures the total factor costs of the goods and services produced by the economy." In other words, what is paid out for production must necessarily equal the income of those who share, directly or indirectly, in the process of producing. For example, total wages are the income of labor; but if we think of the cost of the product, they are also the total cost of labor.

National Product

The income received in 1939 is listed on a previous page. A substantially different set of figures can be consulted to learn what was spent for the current product. Since what was spent was used to buy the goods and services produced, these figures can be regarded as the money value of the national product.

Individual consumers spend money for food, clothing, rent, and a thousand other things. These expenditures can be added up by consulting the statistics of sales.

Business also spends money. In order to avoid duplication, we cannot count what it pays out for making the goods which consumers buy. That is all included in the item in the previous paragraph. What we must count is that part of business profits and borrowings which is spent directly by business itself for new investment in such things as buildings and machinery. For convenience' sake, all the new houses built are included in business spending, whether they are paid for by investors or by the individuals who live in them.

Some investment goes to foreign countries. It is convenient to make a separate item for foreign investment, since this can be checked by a separate group of figures about foreign trade and other international transactions.

The spending of individuals and business listed above does not include what these groups pay in taxes or lend to government by buying government bonds. It is better to regard government (federal, state, and local) as a sort of buying agency of the people, and to list separately what it pays out for the services of government employees and for goods.

There are, therefore, four main items of expenditure (or product) : (1) those goods and services bought for personal consumption; (2) those goods which business bought for investment within this country; (3) investment abroad; and (4) those goods and services that government bought. These are the four great "factors" of the national product which together determine how large that product is, when measured by what is spent for it. It is convenient to list them separately in analyzing the total demand.

In 1939 the figures were:

Gross National Product (1939) Measured by Expenditure	(Millions of Dollars)
Individuals spent for consumption	67,466
Total gross private investment in the U. S. was	9,004
Foreign trade and other transactions resulted in a net increase of foreign investments amounting to	888
Finally, the federal, state and local governments bought goods and services for a total of	13,068
These various uses of money accounted for everything produced and sold for final use. This total is called *gross national product*	90,426

Why is this total called "*gross* national product"? In the definitions previously quoted, the terms used have been "*net* income" or "*net* product."

The difference is that in the figure here given for private investment in the United States, no allowance is made for the wearing out or junking of old machinery or buildings during the year. In other words, there is no allowance for depreciation or obsolescence. Charges for these items are almost invariably made on their books by business concerns, and are deducted before profits are reckoned, or any use of profits is made. Gross national product is therefore larger than net national income.

The figure for gross national product is useful in showing how much was actually spent or produced in a given year. Net product or net income is a better figure, however, to indicate how much material progress is made over a series of years.

In order to change gross product to net product, it is necessary to make a deduction for depreciation which will show what "net investment" was. This deduction is called "capital consumption allowances." In 1939 it amounted to 8101 million dollars. If this figure be subtracted from the gross national product, there is left a total which is called the "net national product," amounting to 82,325 million dollars. According to definition, income equals product. But even the figure called net national product is considerably larger than the figure on page 21 for net national income. The difference is not a result of errors in counting, adding, or subtracting. It now becomes necessary to see what does explain it.

Double-Entry Check

Double-entry bookkeeping is familiar in every business. It means merely that every transaction is entered on both sides of the books. Thus, money received for goods goes on one side, and the value of the goods sold on the other. The totals must balance if the books are accurately kept. A similar procedure now enables a check on national income accounting. Income received was totaled in the first table. From largely independent figures, we have listed how the money was spent, or otherwise disposed of. But still the books do not balance.

To discover where the discrepancy lies, it is necessary to check the figures for expenditure against the figures for income, to see whether there is any kind of spending listed for which no corresponding source of income is included.

First, personal expenditures for consumption come out of personal incomes, all listed in the first table. No discrepancy there.

Second, investment, both domestic and foreign, comes partly out of personal incomes and partly out of business profits, both accounted for in income received. Again no discrepancy.

Government spending for goods and services is paid for partly by taxes and borrowings which come out of personal incomes. All personal incomes are included in the first table. Government spending also is paid for in part by taxes and borrowings derived from business. Business *income* taxes and bond purchases come out of business profits, which are listed on the income table. But business also pays many other kinds of taxes—property, tariff duties, etc., which are charged up to expenses. Business expenses are not included in the income table.

An allowance must be made, in our double-entry bookkeeping, for these *indirect* and other business taxes. In 1939 they amounted to 9365 million dollars.

Government-interest payments have been excluded from both income and expenditure. So have "transfer payments"—that is, receipts and expenditures for social security and the like.

Two more minor technical corrections must be made, called "business transfer payments" and "subsidies minus current surplus of government enterprises." Then we arrive at comparable totals for income and for expenditures—or product, which is what the expenditures buy. There remains a difference of 462 million dollars, which represents some unknown inaccuracy in the compilation of the figures, called "statistical discrepancy." It is probably caused in part by the fact that many incomes are under-reported in income-tax returns, which are used in compiling the figures. This is a small error in so large a total—less than half of one per cent.

The way in which income is listed on the left-hand side of the table on page 28 is usually called "Income by distributive shares," since it shows what is paid for labor or personal services and what for ownership of property in rent, profits, and interest. Income received by unincorporated enterprises of course includes some of both kinds; a farmer, a storekeeper, or a doctor usually does not pay himself a salary, yet his profit includes remuneration for his work. One can arbitrarily divide this item between labor income and property income, on the basis of estimates.

Another way of classifying income is by the source—that is, the industry or occupation from which it is derived, such as manufacturing industries, mining, construction, agriculture, trade, finance, government.

Many kinds of subdivision or cross-classification are possible, each with its own uses.

Spending or product also may be broken down to yield a large amount of valuable information. The classification used on the right-hand side of the table covers expenditures made in any year for (1) consumer goods and services; (2) private capital; (3) foreign investment; and (4) government.

The following table summarizes these figures for 1939.

National Income and Product Account, 1939[1]
(MILLIONS OF DOLLARS)

Income		Product (Measured by Expenditure)	
Compensation of Employees		Personal Consumption	
Wages and salaries	45,745	Expenditures	67,466
Supplements	2,075	Gross Private Domestic	
Income of Unincorporated Enterprises *	11,282	Investment	9,004
Rental Income of		Net Foreign	
Persons	3,465	Investment	888
Corporate Profits		Government Purchases of	
Dividends	3,796	Goods and Services	13,068
Undistributed profits	1,209		
Profits tax liability	1,462		
Inventory valuation adjustment	−714		
Net Interest Received by Persons	4,212		
	———		
National Income	72,532		
Indirect Business Taxes	9,365		
Business Transfer Payments	451		
Less: Subsidies minus current surplus of government enterprises	−485		
Statistical Discrepancy	462		
	———		
Net National Product	82,325		
Capital Consumption Allowances	8,101		
	———		———
Gross National Product	90,426	Gross National Product	90,426

* Including inventory valuation adjustment.

Another important way of classifying national product, which cuts across this one, is that which distinguishes durable goods, semi-durable goods, and perishable goods, or, lumping the last two items together, simply durable goods and non-durable goods. Automobiles, furniture,

[1] Compiled by the U. S. Department of Commerce (from its publication National Income, Supplement to Survey of Current Business, July 1947, p. 2).

household equipment, are examples of durable goods. Clothing is a semi-durable. Food is classified as perishable.

There are also various subordinate accounts which show what happens in parts of the total picture. There is, for instance, a consolidated business account, which shows what business, if considered as a single concern, pays out and what it receives. The items in this table are almost the same as those for business in the national account.

There is, besides, a personal income account, which shows the total of individual incomes—this time including governmental interest—and, on the other side, personal expenditures, tax payments, and savings. An important term which will often be encountered in economic discussion is "disposable income." This is the sum of all personal incomes, less what individuals pay in taxes. It is usually employed when discussing the effect of income on markets for consumers' goods, or the division of personal incomes between spending and saving.

Two other subsidiary accounts are separated out by the Department of Commerce. One is the government account (including, of course, federal, state and local governments). The other is the foreign account. It is worth pausing a moment on the foreign account to see why, in our general table, only the *net* foreign investment (or its opposite, disinvestment) is included as a separate item in the list of spending.

The proceeds of foreign trade activity, in so far as they augment incomes, turn up in wages, salaries, profits, interest, or rent, listed on the income table. There is no need of a separate classification for them. What Americans buy from abroad is automatically included in the various items of expenditure. We need include only the *difference* between incoming and outgoing payments. That difference must be an increase or decrease in debt.

The debt of the debtor is an investment of the creditor. There is no way other than by loans (or gifts) to finance an export surplus or an import surplus—unless payment is made in gold. By including net gold shipments in the figure for net foreign investment, we cover the whole matter.

It is necessary to add to the table this item for net foreign investment in order to balance the books, since what a country pays for imports in any one year almost never balances what it receives for exports.

Real National Income

The fact that the national income must be stated in money terms introduces distortions in comparing one year with another, since prices change, and dollars will buy more of the same goods and services at some times than at others. When no corrections are made for price changes in tables of the national income covering a series of years, there is usually a note that the totals are in terms of "current dollars." It is possible to correct for price changes, however, and this is usually done. When the correction has been made, and the yearly totals are still stated in the form of dollars, the note will state "1939 dollars" or "1929 dollars," or whatever year happens to be chosen as the basis of value. Income after correction for prices is usually spoken of as "real" income, whereas without correction for prices, it is called "nominal" or "money" income.

A convenient way of showing changes over a period of time is by the use of index numbers. When this is done, the income of a certain year (or average of years) is given the number 100, and every other year in the series is expressed as a percentage of the base. Thus, if 1929 be chosen as the base, or 100, and income in some other year is 75 per cent of the base year, it will be represented by the number 75. Tables of index numbers usually carry a note indicating the base year, such as (Base —1929) or (1929=100).

From this brief review of some of the more important ways of reckoning or classifying figures of the national income, it will be seen that an enormous amount of information of many sorts may be derived from them. The task of revising and completing the computations is still far from finished. Because of the gaps in the raw data and the estimates which play a large part in the process, it is risky to place too much reliance on any exact accuracy in the figures. Yet they do tell us much that is significant, as succeeding chapters will show.

3

The National Income as History

THE GREAT mass of information accumulated in building up the totals called the national income can be used for dozens of different purposes. One of the most interesting is to learn what has happened to American economic fortunes in the past. Such facts also have a bearing on the present and the future. Some of them show trends which we hope will continue; some depict conditions which we may want to change, or misfortunes the repetition of which we should like to avoid.

Simon Kuznets, who did much of the pioneer work on the subject, provided a convenient summary up to World War II in his book *National Income, A Summary of Findings*. The figures he uses are not quite the same as those now published by the U. S. Department of Commerce, since the latter have embodied minor corrections and are based on slightly different definitions. These differences are not important for most purposes, however, and for the general conclusions in this chapter, Kuznets' figures are used. Substitution of the government figures would not substantially alter them.

It did not require study of the figures to tell Americans that there had been a remarkable increase in incomes since the Civil War. The figures, however, not only confirm this impression but add definiteness to it.

Per Capita Income

To give meaning to such a demonstration, it is necessary to eliminate the effect of price fluctuations, since a given income could purchase more at some times than at others. This is done by stating the income for all years in dollars having the same purchasing power as in 1929. The yearly income of the population, if divided evenly, would have averaged $215 a person for the decade 1869 to 1878. In 1919 to 1928, it would have averaged $612, or almost three times as large as fifty years before. For the depression decade of the 1930's, the income per person fell for the first time in the history of the United States, at least since the Civil War. It was reduced to a yearly average of $572. Bear in mind that this represents not just the dollars received, but what they could buy.

Increase of income does not necessarily mean an equivalent increase in welfare or in satisfaction in life. At the time of the Civil War, for instance, more of the people lived on farms and ate food that was raised on the farm instead of buying it. Home-cooked food consumed on the farm does not appear in the income figures. The same is true of clothing made in the home in both city and country, laundry or other work done without pay by the women of the household, a much larger propotion of whom now have paying jobs. Yet in so far as welfare depends on buying, people on the average have certainly grown more prosperous.

Shifting Occupations

How was income apportioned among various important occupations? Here we find that marked changes have taken place as the years have rolled by. The share of agriculture has steadily declined. In the decade 1869 to 1878, agriculture had 27.5 per cent of the national income against only 10.5 per cent in the decade 1919 to 1928. During the same years, the share of manufacturing increased from 17.1 per cent to 21.9 per cent, and mining increased from 1 per cent to 2.5 per cent. Construction was variable, but had a smaller share of the national income at the end than at the beginning.

Many people assume that the relative shrinkage of agriculture was caused by a corresponding expansion of industry. The figures show that this was not wholly the case. While the share of manufacturing in the national income did increase, it did not grow nearly as rapidly as the share of agriculture fell. There has been a steady drop in the relative importance of commodity production taken as a whole. Those engaged in producing goods on farms, in factories, and in mines had 51.6 per cent of the national income in 1869 to 1878 and only 39.3 per cent in 1919 to 1928. The reason is an important advance in efficiency of production, which has made possible a great expansion in wholesale and retail trade, in finance, and in professional and other service occupations, such as teaching, automobile service, hairdressing, and so forth.

The shifts in shares of the national income which have occurred from decade to decade were accompanied by roughly corresponding changes in the number employed in each major occupation. After 1928 the number engaged in agriculture continued to fall off, not only relative to the total population but absolutely. The relative growth of manufacturing ended immediately after World War I. From 1919 to 1928 its share of the total national income decreased slightly, while the number employed was about the same as in the preceding decade. In the thirties manufacturing employment also had not only a relative but an absolute decline.

Some Groups Earn More Than Others

We all know in a general way that average incomes in some industrial groups are higher than those in others. This has been true for many years. In the period between the World Wars, the per capita income of workers in agriculture was just about half the per capita income of workers in the nation as a whole. This does not count the food and housing which they received without paying for it. Per capita income in mining and manufacturing was at the average level. The per capita income of those engaged in trade was 10 per cent above the national average; in transportation and public utilities, it was 20 per cent above; in construction, it was 30 per cent above. At the top of the list stood finance and real estate, with 40 per cent above the average, and government employment at the same figure. These figures refer to payments for work, and do not include profits, rent, or interest. It is interesting to note, too, that the lowest per capita returns were received by those who occupy a stage in production which is farthest from the ultimate consumer. Those engaged in primary production receive less than those engaged in secondary production. These, in turn, receive less than those engaged in transportation and distribution.

Inequalities of Distribution

Most of us are interested in the inequality of the distribution of income by size. There are no continuous yearly figures on this subject except for the very highest brackets. In the twenty-year period from 1919 to 1939, the upper one per cent of the income receivers had 13.7 per cent of the national income. The upper 5 per cent had 26.8 per cent of the income, and the lower 95 per cent had only 73.2 per cent of the income. Many people believe that high income-tax rates in the upper brackets greatly modified this inequality. This was not true, since if income taxes are taken into consideration, it makes a difference of less than 1 percentage point in the share of the total received by the upper 1 per cent or by the upper 5 per cent of the income receivers.

A more detailed study of income distribution was

made for the year 1941 by the government. In this prosperous period of defense production, 16 per cent of the families in the nation had an income of less than $500 each, and received only 2 per cent of the total income. Another 18 per cent of the families had between $500 and $1000 each and received only 7 per cent of the total. Sixty-four per cent of the families in the nation had an annual income of less than $2000 and received less than a third of the national income. Only 4 per cent of the families had over $5000, and these families had 22 per cent of the income. These same figures show that the lowest income groups spent more than they received in income, apparently by running into debt or on account of relief payments. Families getting between $3000 and $5000 apiece spent as much as they received. *The only personal net savings shown by the table arise in the income group receiving over $5000.* This does not mean that nobody with an income of $5000 or less saved anything, but the deficits in this bracket balanced the savings.

It is clear from these estimates that the distribution of income was extremely unequal. The popular impression that great inequality exists is correct. It is equally clear, however, that the poor could not have been greatly benefited by distributing among them all the surplus income of the rich. If we had taken away from all the families having incomes of $5000 or more per year all their income above $5000, we should have had only 12 per cent of the total national income to distribute among the 96 per cent of the families who received less than $5000. As a matter of fact, we could not safely redistribute as much as this, since much of it goes into saving which is necessary for the expansion of productive facilities. If the sums necessary for new investment were deducted, there probably would not have been more than 5 per cent left for redistribution. This is not necessarily an argument against more equal distribution, but it indicates that the main objective must be production of more to be divided.

Where Incomes Come From

As would be expected, the lower income groups receive most of their compensation as employees or for services rendered, whereas the upper income groups derive more from dividends or property income. In the inter-war period, 69.7 per cent of the dividends went to the upper 1 per cent of the individuals receiving incomes. Rent and interest are somewhat more equally divided, but not wholly so. Taking the national income as a whole, and classifying it according to the various types of payment, we find that 84.2 per cent was paid for work performed, in the form of wages, salaries, or earnings of working owners such as farmers, while the other 15.8 per cent was paid as a return on property. Further subdivision shows that 66 per cent of the total national income consisted of employee compensation, 18.2 per cent went to individual businessmen and farmers, 5.9 per cent was paid in dividends, 6.9 per cent in interest, and 3 per cent in rent.

What Consumers Bought

Many other important and interesting facts are known about the distribution of income, such as the differences among regions and the differences among cities of various sizes, rural and farm localities. But let us pass on to a few things that the figures tell us about purchases by consumers. In the inter-war period consumers spent about 40 per cent of their budgets for perishable goods such as food, about 15 per cent for semi-durable goods like clothing, slightly less than 10 per cent for durable goods such as automobiles and furniture, and 36 per cent for services. The chief change during the period was that a smaller proportion was spent for perishable goods and a larger proportion for durable goods and for services. This trend has continued.

It is a very important fact that as incomes grow, people spend a larger and larger proportion for things like automobiles, electric refrigerators, and furniture. These things can be made to last longer without replacement, at a pinch. Their purchase is more largely financed by credit. Therefore the demand for them may vary widely from one year to another—much more widely than the demand for food or even clothing. This means that as they assume a more prominent role in the economic picture, the instability of the whole order, and of the employment and income which it supplies, is greatly increased. A falling off in production of motorcars, for instance, can easily occur and can have serious consequences.

How Much Was Saved

Between 1919 and 1939 about 6 per cent of the national income was saved and invested in new capital facilities. This proportion of saving to total income seems surprisingly small when we consider the immense increase in production and productive efficiency which occurred, especially in the first post-war decade. We must remember, however, that during the thirties, especially while depression was rampant, new investment was small, and in some of these years net capital formation was a negative quantity.

Nevertheless, we are justified in drawing the conclusion that a rather small amount of saving is all that is necessary to buy the equipment for greatly expanded production. Consequently, if the much larger national income which has resulted from the recent war is all to be spent, there must continue to be a great increase over the pre-war days in the consumption of the population.

After the post-war boom, there will be no incentive for even the usual new investment unless this expansion of consumption continues. Of course, if people greatly decrease their spending, either in consumption or in the purchase of capital goods, the total income itself will decrease and we shall have a depression.

The new figures of the Department of Commerce differ from those compiled by Kuznets in estimating the amount of personal savings more than in any other item. But these new figures indicate even smaller savings than Kuznets calculated. They therefore reinforce the point that we cannot keep the post-war productive capacity busy without extremely large consumption, and a relatively small proportion need be saved in order to increase the productive plant of the nation rapidly.

Losses From Depression

The national income figures also tell us a great deal about the huge losses incurred by the nation because of depressions. In the two decades between the wars there were five years which showed large declines in the national income and three others in which the increase was negligible. The same is true of employment. The average drop in income for each depression year was 4 per cent of the average yearly income for the whole period. The loss occasioned by depression is still more dramatically shown by another type of calculation. Kuznets has calculated what the national income might reasonably have been expected to be during the inter-war period if it had continued at the level of full capacity. This estimate is a conservative one since it assumes that the production achieved at the top of prosperity was somewhat larger than could have been steadily maintained. On this basis, he finds that the income might have been more than 13 per cent greater for the twenty years than it actually was. Remember that this calculation includes not only the depressed thirties but the prosperous twenties.

During the Great Depression of the 1930's, total income payments to individuals dropped almost half—by 45 per cent. A study of family incomes made by a sampling method shows an average fall of 37 per cent between 1929 and 1933.

The depression also had curious effects on the distribution of income. Inequality was, on the whole, increased. The gulf between the lower and the upper groups of income was widened. Within the lower 50 to 70 per cent of incomes, there was also more inequality. This is probably accounted for by the fact that the millions of unemployed received comparatively little, while those who held their jobs were not much worse off than before.

Within the upper income groups which received a larger share of the total income during the depression, there was less inequality than usual. The very topmost incomes declined relative to the others. These high incomes are derived largely from profits, which fall drastically during depression.

Conclusions for Policy

What broad conclusions are we justified in drawing from the facts here summarized? First, while it would be socially desirable to reduce the extreme inequality in the distribution of income, any progress in this direction would not help much to alleviate poverty without large increases in the total product of the nation. This conclusion is reinforced by the experience of inflation after the war. Though in 1947 almost everybody was employed, and the total output of the nation was about half again as large as before the war, there still were not nearly enough goods to satisfy the demands of the population. When people have plenty of money to spend, it becomes more apparent that not enough is produced, even yet.

Second, there will always be plenty of opportunity to sell to consumers all the extra goods that might be produced if we can distribute to those at the lower end of the income scale sufficient money to buy them. It is obvious that the 64 per cent of the families who had incomes of less than $2000 in 1941 must have had many unsatisfied wants. We are nowhere near the saturation point of people's desire for goods and services.

Clearly, the principal economic objective must be to increase the production and distribution of needed goods and services. That this objective is attainable is indicated by the fact that there has been a marked tendency to increase over a long period of years. Whether that increase has been greater or less than it might have been under some other form of economic organization, these figures do not indicate.

Finally, the most important obstacle to the desired increase which the figures do show is the frequent recurrence of depression and unemployment. Avoiding depression, therefore, is a most important objective.

There is nothing startling about these conclusions, which are probably accepted by the majority of Americans today. It is worth remembering, however, that they were not nearly so generally accepted within the memory of many living men. The experience of the Great Depression, of course, had much to do with the change in the mental climate, but the advance in economic science, which is marked by the study of the national income and which dates from the end of World War I, has given us the solid documentation which has established the character of our situation beyond dispute.

4

The Gains in Production

So FAR, this book has been concerned with income reckoned in dollars. But income also means what is physically produced. It is because production has grown markedly in the United States that income has increased. There is a more direct way of visualizing the increase of production which is the basis of material welfare— or, at least, a way which seems more direct. This is to measure the growth of the physical product itself, a growth which can be stated in percentages or index numbers without any reference to dollars at all.

Careful studies of this subject have been made by the National Bureau of Economic Research and others; figures of production are now currently published by the government. It has become a commonplace that these figures show a tremendous advance.

The Over-All Growth of Production

The reliable studies of production cover agriculture, mining, manufacturing, gas and electric utilities, and steam railroads. These industries account for a little less than half of the national income; they cover nine-tenths of the output of goods in the country, but only one-tenth of the services. They employed two-thirds of the labor force in 1899; almost half in 1939.

Between 1899 and 1939 physical output nearly tripled. It would have grown much more than this except for the depression of the 1930's. During World War II industrial output more than doubled again; in 1947 it was about 85 per cent more than in 1936-39. This more recent figure covers only manufacturing and mining; there was also a substantial increase in the output of agriculture, railroads, and utilities. It is estimated that the national product as a whole is more than 50 per cent above the pre-war level.

Aside from the fact that physical output figures do not include anything like the whole of our economy, they have two important defects. First, they do not measure improvements in quality. One automobile produced in 1909 counts just as much in the figures as one produced in 1939, yet the second provided much more transportation and superior comfort and cost less to keep on the road. In so far as products have been improved, the advance is greater than is shown in the figures. The second important defect is that things done at home or on the farm for direct consumption are not counted. There has been a long-continued tendency for people to buy more food and clothing and make less for themselves, to send the laundry out, and to use gas or oil instead of chopping wood. In so far as this happens, the figures overstate the advance. Experts believe that the first tendency outweighs the second, and therefore, that the progress in production has been greater than is shown in the statistics.

The average *rate* of advance in the thirty years from 1899 to 1929 was 3.5 per cent a year. Then production declined, and by 1939 had barely recovered to the level of ten years earlier. Another big boost was occasioned by World War II. It is no surprise to learn that depression decreases production and war is likely to raise it, but the problems involved in these facts are emphasized by the actual figures.

What interests us is not the mere growth in the total product, but the relation between that total and the

number of people in the country. No benefit is possible if
population triples while production triples. But up to
1929 nothing like that occurred. Not only did physical
output increase, but it increased so fast that in 1929 there
was 74 per cent more for each person in the country than
in 1899. The depression reversed this process. In 1939, the
output per head was less than in 1929, though still 65
per cent higher than at the beginning of the century. It
jumped again during the recent war.

Production Per Worker

A highly significant fact is that as time has gone on,
more and more has been turned out in relation to each
worker employed. In 1939 the number of workers making
these physical goods was only a third larger than in 1899,
while the output itself was almost three times as large.
The result was an increase in output per worker of 122
per cent. Here are the figures, taking 1899 as the base, or
100.

	1899	1909	1919	1929	1939
Output	100	146	195	283	289
Employment	100	129	153	150	130
Output per Worker	100	113	127	189	222

This increase is sufficiently striking, but what is even
more so is that while the number employed increased by
one-third, the average hours worked decreased by a third.
The consequence is that *in 1939 each man-hour of work
produced nearly three times as much as in 1899.*

This advance was not interrupted by the depression, as
was the total produced and the total employed. Indeed,
the years 1937 to 1939 showed one of the most rapid increases in output per worker on record, or 5.4 per cent a
year. The rate of increase does show a variation from time
to time. Sometimes it temporarily slows down to almost
nothing. Apparently, there was little or no increase in

output per worker during World War I, and some believe
that this was true also in World War II. Millions of inex-
perienced people were hired; new processes had to be
learned; there were interruptions in the flow of materials.
After World War I, especially when the 1920 inflation
had come to an end, the advance was very rapid and more
than made up for lost time. This may now happen again.

It is fruitless to discuss how much of the improvement
in output per worker or per man-hour is due to harder
or more skillful work, how much to advances in tech-
nology, how much to additional capital equipment. All
have contributed, but the relative contribution of each is
difficult to measure. Our working population has become
better educated, is far more familiar with machine
processes than in earlier times or in more backward
civilizations, is held to higher standards of output, and is
given incentives by skillful managements. If this were all,
we should be less hopeful of continual advance in the
future, since there is a limit to the intensification of
human effort. Technological advances also have made
tremendous contributions; they are the marvel of our age
and the end is not yet in sight.

So far, one factor has been of supreme importance—
the substitution of mechanical power derived from coal,
oil, gasoline, gas, or falling water for the muscles of men
and animals. The application of such power just about
parallels the advance in productivity, as the increased
output per unit of work is called.

Service Output

The statisticians have not yet got around to measuring
the output and productivity of many service industries.
Service industries include railroads and utilities (which
have been measured) and also trade, finance, real estate,
professional service, domestic service, government service,
barbers, beauty shops, and a host of other occupations.

The difficulty of obtaining figures for physical production in such occupations is obviously great, though perhaps not insuperable.

It is highly probable that in many of these cases there has been little or no increase of output per person employed; certainly not so great an increase as in the industries making physical goods, where machines and production lines are prevalent. Nor is it clear that in services we want greater productivity of a type that could readily be measured. If, for instance, the number of pupils per teacher were doubled, we might get a higher productivity of school graduates, but we certainly should not get better education. Automats have fewer employees per calorie of food, but those who can afford the time and money usually prefer restaurants with individual service.

A striking characteristic of the growth of our order is that as physical productivity increases in the industries producing material goods, a smaller and smaller proportion of the population is employed in them, while a larger proportion is employed in the service industries. More than half of the labor force is now engaged in providing services. The chances are that this shift will continue. Because service industries, by and large, are not so well adapted to use of machinery and power and so cannot enjoy rapid increases in productivity, the total gains in output are likely to slow down as time goes on.

Growing and Declining Industries

The increases in total production and in production per man-hour which are shown by the general averages do not mean that every industry is gaining, or that those which do grow gain at the same rate. There are always some industries which grow very rapidly; others that decline. Indeed, the figures seem to show that the more rapid is the growth of the general average of production, the more declining industries there are. New articles take the place of old ones: in 1919 few private citizens had radios; by the end of the decade, almost every family had one, but meanwhile the output of pianos had shrunk almost to the vanishing point. The great growth of the automobile industry in the first three decades of the century was paralleled by a drop in the production of carriages from nine hundred thousand in 1899 to thirty-six hundred in 1929.

Less spectacular than contrasts such as these, but equally important, is the fact that every great industry seems over a long period to follow a typical path of growth ("growth curve" to economists), which shoots up rapidly at the beginning, then slows down and tends to flatten out, if it does not drop. This may not be because anyone wants to substitute new products for the old one, but merely because eventually the industry approaches the limit of its possible market, the number of the consumers and their incomes being what they are.

Look at the big and important industries at any particular time, and it will usually appear that their growth is slowing down. That was the case of the automobile industry toward the end of the 1920's, and it will probably be true of it again as soon as the shortage caused by lack of production during the war has been made up. It has been true of the railroads ever since 1920, although for nearly a century before that railroad building had been one of the chief industries of the country, giving employment to hundreds of thousands and providing large markets for steel, lumber, and other materials.

Some economists believe that new and expanding industries will always come along to take up the slack left by those which have reached the tops of their curves. Others think the country is approaching "maturity," and can see no new industries of sufficient importance to replace the decline of the old ones. In that case, the total advance of production will slow down, and we are likely to suffer unemployment, unless something is done to increase sufficiently the consumption of goods with which we are already familiar, or the government starts a lot of new and useful projects.

Still another theory holds that major swings of prosperity or depression are caused by the fact that though great new industries do appear from time to time, they do so only spasmodically. The spurts of construction and expansion, which in turn are largely dependent on the application of new inventions or discoveries, bring years when the trend is upward; after that we may have to wait more years before another such stimulus appears. Whether or not this theory is true, there certainly have been periods when growth has been faster and other periods when it has been slower.

Output Per Unit of Capital

It is largely irrelevant whether those who do the work or those who supply the money, the buildings, and machinery "deserve" the greater credit of increases in production. The question of how income ought to be distributed does depend in part upon the necessity of giving the proper reward or incentive for the contribution anyone may make to the creation of that income. Yet when the growth of the total product is under examination, it is beside the point to analyze the question of who has contributed most and what his compensation justly ought to be.

In another sense, however, the contribution of "capital" is of the utmost importance to understanding the tendency of production. It is possible to conceive of a situation in which, on the basis of statistics showing merely the increase in production per worker, it would look as if great advances in popular welfare were being made, when, as a matter of fact, individual consumers were not being benefited at all. This would be the case if so much labor and materials were being diverted to the production of more efficient machinery that the output of goods for consumers was not increasing any more rapidly than the population.

Just to make the point clear, let us use a ridiculously simple illustration. Suppose there were only a hundred workers in the country, all making things for consumers. Then let us suppose that someone devised a new invention by the use of which twice as many consumers' goods as before could be turned out with the same amount of labor. Would all of the hundred workers now produce two units of goods for us in place of every one they had previously made? By no means. Some of the workers would have to be kept busy making the new machines or digging the materials for them out of the ground. Suppose it took twenty-five workers to do this. We should then have only seventy-five workers left to make the consumers' goods; they would, using the new machinery, turn out only half as much again as was produced before. Consumers, including all workers, managers, capitalists, and others in our society, would derive less benefit than the increase in output per worker.

Such an extreme case as this seldom happens, yet the illustration shows us why it is important not to forget the efficiency of what we call "capital." The cost of labor effort and materials which go into productive buildings and machinery is difficult to measure, but the book value of productive equipment certainly has some relation to what has had to be paid to bring them into being. An extremely rough measure of this kind, made by George I. Stigler for the National Bureau of Economic Research, indicates that the growth in output per unit of capital in manufacturing industries is in many cases considerably less than the growth in output per worker. This would be expected, since the cost of the new plant and equipment needed to turn out more output per worker often increases very rapidly as the years pass (entirely aside from general price fluctuations) .

For instance, according to these figures, the output per worker in printing and publishing grew 150 per cent between 1904 and 1937, while the output per unit of capital in the same industry increased 142 per cent. Greater contrast is revealed in the making of transportation equipment, in which automobiles bulk large. Here the output per worker grew in the same period 308 per cent, but the output per unit of capital increased only 130 per cent. In a few cases, like petroleum and coal products, the output per unit of capital actually fell, because of the great expense of modern refining plants.

It is clear that in all these cases the growth in over-all efficiency of production is somewhat less than the increase in output per worker. Exactly what it is depends upon how much of the gain is attributed to labor and how much to capital. Sometime all this may be worked out more precisely; for the present, it is enough to say that when we gain more output per labor-hour by the building of better plants and machinery, the resources of labor and materials which are used up to create those plants must not be forgotten in reckoning the benefit to society.

Production and the Future

The long-term consequences of the gains in productivity are of staggering importance. It is due to the progress hitherto made by the United States in this regard, for instance, that with only one-fifteenth of the population of the world and about the same proportion of its land area and resources, we turn out more than half the industrial production of the whole globe. If we go on advancing at the same rate, it will not be many decades before everyone in the nation can have decent housing, adequate nourishment, a good education—in short, the material essentials of a good life. If other nations catch up with us, the transformation of human life among the vast populations below the poverty line could be beyond reckoning.

A century and a half ago Malthus advanced the theory that the enormous growth of population would force almost everyone to live at the lowest level upon which life could be sustained. Nations like China or India have in the past seemed to exemplify his theory; only the great increase in productivity in the Western world, combined with a slowing down of the birth rate, has upset it. Yet it still might come true if these tendencies should be reversed.

A more comfortable prospect is that as time goes on our demands for things may be satiated, and we shall rather want more time for recreation and for aesthetic or intellectual pursuits which have no commercial value. Already working hours have been drastically reduced. Sometimes people worry about "overproduction" in the sense that they believe we can make, or shall soon be able to make, more things than people want. So far this has never happened, except in the case of a few specific persons or isolated products. What has happened is that at times more has been made than people in general were able to buy, which is a different matter altogether. If everyone's wants for goods really were satiated, there would be no need to worry about unemployment, since an unemployed person who had all the food, clothing and everything else he wanted would be indistinguishable from a gentleman of leisure.

5

Ups and Downs of Business

NOTHING REVEALS the myopia of classical economic theory
for the facts of life more clearly than that its system of
thought allowed no place for industrial depressions. One
would think that men who set out to explain economic
phenomena would study the oscillations of production
and employment almost before they noticed anything
else. What early theorists did, on the contrary, was to
conclude that the natural forces of the economic order
would keep it in balance at a high level of employment.
There could, they believed, be no such thing as a general
shortage of demand.

Later theorists of the classical school could not deny
that unemployment did occur or that business conditions
were sometimes bad. These inconvenient facts did not,
however, induce them to alter the fundamentals of their
thought. They found two kinds of excuses for the wide
difference between the actual behavior of the world and
the manner in which their theory said it ought to be-
have. One was that something from outside the economic
order intervened and set it wobbling. For instance,
Stanley Jevons suggested that business fluctuations arose
from variations in the farmers' income, that these, in
turn depended on weather, and that there was a weather
cycle arising from sunspots. This was a pretty reasonable
guess, as guesses go. There is indeed a weather cycle. Yet
there does not seem to be any regular relation between

the farmers' fortunes and the oscillation of business conditions.

The other road of escape which the classical theorists found was to admit (and usually to deplore) that people did not act according to the premises on which their theory was built. Monopolies or partial monopolies interfered with perfect competition. Capital did not flow easily from one use to another; labor would not readily shift jobs. Prices, instead of responding promptly to changes in demand and supply, were "sticky." These frictions and rigidities delayed the automatic readjustment in the direction of wholesome equilibrium which the theoretical system expected.

There was nothing wrong with these observations; such rigidities and frictions do indeed exist. What was wrong was the fact that many followers of the theory stuck to their premises and attempted to brush away the facts as something extraordinary, outside the "natural" order. The result was that their only remedy was to exhort people to comply with the "natural" law. This is shown no more clearly than in the classical anaylsis of unemployment. According to the doctrine, employers would always hire everybody seeking work if only wages were low enough. Falling prices were supposed to stimulate demand; falling wages would therefore stimulate the demand for labor. The cause of unemployment was thus found in the refusal of workers to accept low enough wages. This reasoning led to the strange conclusion that unemployment was the fault of the unemployed; there was no such thing as involuntary employment. This reasoning did *not* lead to any scientific discoveries about depressions.

As more attention began to be paid to the problem, a host of explanations was offered for the persistent rhythms of business activity. It would require a long book to summarize the many varieties of business-cycle theory. But such a book would be more confusing than enlightening, since most of the explanations advanced have been based on guesses, partial observations, emphasis on one or two factors which overlooked the rest. Many of the theories are mutually inconsistent; some of the better ones fit some of the facts, but none fits all of them.

Scientific Exploration of the Cycle

At last a serious effort was made to accumulate and analyze all the facts available. Ever since the 1920's the National Bureau of Economic Research, with a competent staff of economists and statisticians, has been studying the data under the direction of the chief American authority on the subject, Wesley C. Mitchell. It has accumulated and carefully analyzed mountainous piles of statistics going back for about a century, and for four nations—the United States, Britain, France, and Germany. Its object has not been to prove or disprove any preconceived theory, or even, as yet, to test any hypothesis at all. It has wanted to see first what the business cycle looks like in reality, and how it actually behaves. Once the facts are properly arranged, it becomes easier to formulate appropriate theories and to test them.

One of the first jobs in an effort of this kind was to make a rough definition of just what was the subject under observation. According to Dr. Mitchell's definition, a business cycle is a fluctuation of the aggregate economic activity of a nation; it consists of expansions occurring at about the same time in many lines, followed by contractions. The contractions, in turn, are followed by revivals, which merge into the expansion of the next cycle.

Though there are periods of relative quiet, there never has been a flat calm in the recorded history of any business-enterprise economy. There are longer periods of expansion or contraction, but the business-cycle is defined as the shortest of these up-and-down movements. It is described as "recurrent but not periodic." This means that although it is continually repeated, you cannot derive from its history any mathematical formula by which you might predict when the next boom or depression will arrive or how severe it will be. Some cycles are short and others are long; some are extreme in their rises and falls, others are moderate. Cycles of different duration or amplitude do not follow each other in any regular order. What we do know is that if history is repeated, there will surely be an expansion after a contraction and a contraction after an expansion.

The Method of Measurement

Great care was devoted to devising a method of measurement that would not conceal any assumptions which might need testing against the facts discovered. One favorite method, still prevalent in discussions of the subject, is to assume that there is a long-term trend of business activity which is independent of the shorter fluctuations, and that the business cycle consists of dips below this trend and bulges above it. Use of this method involves charting a trend line and then measuring against it the shorter waves of the business curve.

Tracing the cycle against a trend is unsatisfactory because the measurements of the cycle differ according to how the trend line is placed, and that depends on a rather arbitrary judgment. It was rejected by the Bureau as unscientific. Instead, the Bureau found a way of measuring, charting, and studying each successive cycle without placing it on a trend. This involved no assumption as to what part of a given upward movement, for instance, must be attributed to long-term expansion and what part to cyclical forces. The method was quite clear as to when the top was reached, and when the ensuing decline ended. It showed what happened in between, and permitted a comparison of one cycle with any other.

Measured in this way, there were twenty-six complete

business cycles in the United States between 1836 and 1938. Between 1854 and 1885 the average cycle lasted about five years; since 1885 it has lasted about three and a half. But it would not do to apply this conclusion too literally. This average length covers wide differences. The variation of the length of the cycle within each period is much greater than the variation between the average of the first period and that of later ones. Between 1854 and 1885 the shortest cycle lasted thirty months and the longest one ninety-nine months. In the most recent period the shortest cycle lasted twenty-nine months and the longest sixty-three. Anyone who attemped to forecast the next boom on the basis of the average length of the cycle would be putting his money on a long shot.

Are Depressions Getting Worse?

Karl Marx predicted that the commercial crises which always accompany capitalism would become worse as time goes on. Many, after the disastrous experience of the 1930's, agree with him. The researchers, looking back over the experience of a hundred years, can find no such trend. It is true that the slump beginning in 1929 was unusually severe and unusually long, but there were bad ones in earlier parts of the period studied, in the 1870's and '90's, for example. And there have been some relatively short or mild ones in recent years. The Bureau's judgment on this point is not a mere impression; it is based on the most rigorous and careful statistical analysis.

Lest anyone become too optimistic, however, it is necessary to record the twin judgment that there is no support for the theory, widely held in the 1920's and being revived today, that depressions will get milder as time goes on. Dr. Mitchell finds no factual support for any long-term trend, either for the worse or for the better. He mentions the possibility that a change for the worse occurred about 1914, and admits that it may be true. But our experience since that date does not include enough cycles to warrant a conclusion.

The only change which appears in the statistics—and this is of doubtful validity when the experience of other countries is considered—is an apparent reduction of the average duration of the cycle.

Aids to Forecasting

Any forecaster who bases his predictions on a definite or average length of the cycle, or in any other way assumes that it is periodic in the sense in which movements of astronomical bodies are periodic, should be scientifically suspect. The woods are full of such mechanistic forecasters. Another and perhaps better justified basis of prediction is that based on the belief that some kinds of economic activity customarily turn up or down ahead of others. The studies of the National Bureau confirm that there is a tendency for the turn in the various statistical series to follow in a regular order. Thus, bank clearings outside of New York City usually have already turned upward at the low point of a depression. So have ton-miles of railroad freight, imports and wholesale prices of hide and leather products. Orders for locomotives, freight cars, and passenger cars, bank clearings in New York City, number of shares traded in the New York Stock Exchange, the index of common-stock prices, and plans filed for new buildings in Manhattan have usually turned downward before the top point of a boom.

What is more simple, then, than to look over the current figures and find out whether a change in general business conditions is at hand? It isn't so simple as it seems. For one thing, these leading horses don't always run true to form. Once in a while they will be at the tail instead of out in front, as was the case with the stock market in 1929. Even worse, they often have minor ups and downs during their cyclical expansion or contraction.

The tops or bottoms of their curves can be figured out only after they have occurred. If stocks fall for a month, how can you tell whether they are going on down? Or if they recover after a few weeks, does that mean they are going higher than before, or only that an upward jog has occurred in a downward trend? "One expansion in stock prices lasted eighty-four months, another only five months. . . . In one specific cycle, the standing of freight-car orders at the peak was about 40 per cent above the cycle average, in another almost 500 per cent above."

This does not mean that forecasting is a hopeless task. But it is by no means a mechanical one, to be performed by some simple trick like noting a weather sign. We may hope eventually to achieve perhaps as high a percentage of success as the weather man, and by as complex methods. Good forecasting would involve, first, an inspection of forty or fifty statistical series to make up one's mind where we stood in a given cycle, and then the use of shrewd judgment about other factors.

Are There Other Kinds of Cycles?

Various economists have argued that there are longer cycles of which the short business cycles are like ripples on a tidal wave. Some think there is a ten- or fifteen-year cycle; Kondratieff finds long waves of fifty or sixty years. The Bureau's scientists had to investigate such theories in order to find out whether long trends had any effect on the short cycle. They found no conclusive proof that such longer cycles exist, though there are irregular upward or downward movements lasting longer than three or five years. But whether longer cycles exist or not, they do not affect the nature of the shorter ones in any noticeable way.

There are, as is well known, cycles in specific types of activity which differ in length and timing from the general business cycle. For instance, we find in agriculture a well-defined cycle in the production of hogs, which depends upon the relationship between the price of pork products and the price of corn. There is another cycle in the production of livestock, which has something to do with the time necessary to breed and increase herds and the over-supply that usually follows. There is a very well-defined cycle in building construction which affects many related industries. Its course is ten to fifteen years. These other movements may affect business conditions in general, but they do not seem to have much influence on the special pattern of the business cycle.

The National Bureau does find some basis for one hypothesis regarding a longer "cycle of cycles": "After a severe depression industrial activity rebounds sharply, but speculation does not. The following contraction in business is mild, which leads people to be less cautious. Consequently, in the next two or three cycles, while the cyclical advances become progressively smaller in industrial activity, they become progressively larger in speculative activity. Finally, the speculative boom collapses and a drastic liquidation follows, which ends this cycle of cycles and brings us back to our starting point."

But this hypothesis is advanced only for further study; no simple pattern applies invariably.

The study does substantiate the belief that when the long-term trend of prices is downward, the contraction phase of the cycle is longer, and vice versa.

Facts About the Cycle

The National Bureau has discovered a multitude of interesting facts about the business cycle, some of which have already been published, and more of which will be revealed and discussed in its future volumes. A good many of these facts have a bearing on various theories in the field. Some previous hunches have been corroborated, others have been disproved. Most of these discoveries are too technical to be described in a short treatment of the subject.

My own guess is that the Bureau will never emerge from the statistical depths like a pearl diver bearing aloft that priceless jewel, *the* cause of depressions. It is not likely to say, for instance, that they are caused by over-production, or undercomsumption, or oversaving, or disparity between investment and the propensity to consume. What it will do, rather, is to show in multitudinous detail what the process of business fluctuation is like, and what the relative magnitudes of important factors are. It is likely to cast light on the critical points where intervention might have a desirable effect.

Among the interesting facts made plain—some of which have been observed before, but not with such precision or in the setting of so thorough a view of the whole subject—are the following:

Business cycles affect both prices and volume of production. By and large, they affect any specific industry or occupation by a reciprocal combination of these two. That is, if the price fall of any article is great during a depression, there will be less fall in the volume of its production, and vice versa. Thus, the output of agricultural commodities typically is not much affected by a slump, but the prices of these commodities fall drastically. On the other hand, the output of steel drops far during depression, while the price is much less affected.

That part of investment which consists of business inventories conforms very sensitively to the swings of the cycle. Its fluctuations have accounted for about half the difference in the flow of gross investment between the top of prosperity and the bottom of depression during the years between World War I and World War II. Indeed, they have been so great as to account for more than one-fifth of the cyclical changes in the gross national product.

In general, investment shows wide swings during cycles. Between 1919 and 1935 out of every $100 of gross national

product, $80.80 was spent for consumers' goods and
$19.20 on investment (including replacement of existing
capital). Among the goods consumers buy, their purchases of durable goods, like automobiles and furniture,
show the widest swings. Consumers paid for these $9.50 of
their $80.80 of spendings. Combining the gross invest-
ment (which swings widely) and the purchase of durable
consumer goods (which swings widely), we get $28.70
out of $100, or 29 per cent of the gross national product.
We should have the cycle nearly licked if we could sta-
bilize this 29 per cent (though of course we should have
to be careful, in doing so, not to produce undesirable results of another kind).

Can Depressions Be Abolished?

The scientists now leave this question to be judged
only by inference. Since the business cycle is characteristic
of business enterprise, it could probably be abolished by
substituting some other regime, such as socialism. Within
an enterprise economy, booms and depressions are not
likely to disappear without more radical changes than
have been made during the past hundred years. That
century has seen many tremendous alterations in economic institutions. There has been a great expansion of
industry accompanied by growth of real income and improvement in productivity. Concentration of economic
control has been intensified. Governmental regulations by
the score have been adopted. Cities have increased im-
mensely; population growth has begun to taper off. Statistical knowledge and attempts to predict business conditions have proliferated. The art of business management
has become advanced; accounting, inventory control, and
market studies are widely practiced. There is no evidence
that any of these or other important developments have
made business fluctuations better or worse, or have
greatly changed their nature. How fatuous, in these cir-

cumstances, seems the belief that if only we retain private enterprise without modification we shall be secure!

But the evidence does not prove that the problem is insoluble. Some slight improvements have been achieved. For example, in recent years the swings of interest rates for call money have become smaller because of a deliberate policy by the Federal Reserve System. If the nation as a whole consciously employed measures to moderate other and more important fluctuations, it might succeed, provided it knew enough to adopt the right measures.

Knowledge in this field is rapidly accumulating. Nor need we wait for the final word before doing the best we can on the basis of what is known. The urgent need for a solution, and the attempts to find one in practice, will in their turn contribute to knowledge.

6

Budgeting for the Whole Economy

Advances in ability to control nature are dependent on new tools. Sometimes these tools are mechanical inventions, like the microscope or the spectroscope; sometimes they are tools of thought, like the famous equations of Newton or Einstein. Economists have recently been developing tools of the latter kind, which give promise of real improvement in the ability to analyze what is happening in the economy in such a way that something helpful can be done about it.

The Nation's Economic Budget

One outstanding tool recently invented is a tabulation called "The Nation's Economic Budget." Everyone is familiar with budgets of one kind or another—household budgets, business budgets, the budget of the federal government. This new one is a super-budget covering the activities of the whole national economy—individuals, business, and government. The Nation's Economic Budget could not have been constructed without the work which has gone into understanding and recording the national income. For, like all the budgets for smaller units, it is a matter of matching income against spending.

But there is one important difference between this budget and all the others. A housewife or a businessman, when working out a budget, is concerned mainly with the struggle to make income cover expenses. The aim is to keep income up and expenses down, so that there shall be no deficit and, if possible, a surplus. In the Nation's Economic Budget, however, there cannot possibly be either a deficit or a surplus. Income will necessarily, and without effort on anybody's part, exactly match outgo. If any difference is shown between receipts and expenditures, the economist knows that there is something wrong with the figures. He examines the statistics to find the error. To achieve a balance is merely a matter of checking the accuracy of the calculations; it is not the purpose of drawing up the budget.

This balance is a fundamental equation of economics, like the basic equation of any other science. It is axiomatic. If only an individual were concerned, his income would probably be less or greater than his outgo. The same is true of a business, or of any part of the total economy, such as an industry or the federal government. But if all individuals, all businesses, and all other economic units are taken into account, their aggregate income must equal their outgo, since there is no way for any one of these units to receive income except as a result of equivalent expenditure by one or more of the others.

The purposes to which this equation may conceivably be put are many, but at present there is concern chiefly with examining and forecasting the total of expenditure. What everybody spends constitutes the total demand. This demand in large part determines whether there will be inflation at one extreme or depression and unemployment at the other. It may be possible to predict the future of demand, or even to influence it, if the various units which contribute to the total can be properly classified, and their contribution to demand measured.

The classification used in the table on page 66 is much the same convenient one as on page 28.

The Flow of Income

In order to understand this budget, it is necessary to bear in mind the flow of money about the system. Everybody receives money; everybody pays it out. The table measures this flow at several important gates.

The Nation's Economic Budget, 1946[1]
(BILLIONS OF DOLLARS)

Accounts	Receipts	Expenditures	Excess (+) or Deficit (−)
Consumers:			
Disposal income	158.4		
Expenditures		143.7	
Saving (+)			+ 14.8
Business:			
Undistributed profits and reserves	13.3		
Gross domestic investment:			
New construction		8.5	
Producers' durable equipment		12.4	
Changes in inventories		3.7	
Total		24.6	
Excess of receipts (+) or investment (−)			− 11.3
International:			
Net foreign investment		4.8	
Excess of receipts (+) or investment (−)			− 4.8
Government (federal, state, and local) :			
Cash receipts from the public	56.5		
Cash payments to the public		55.2	
Excess of receipts (+) or payments (−)			+ 1.3
			0

[1] From the Economic Report of the President, January 1947.

It is both convenient and logical to start measuring with consumers since that classification includes every living soul and much of the money flow pours through this gate. We want to measure, first, the purchasing power going to consumers, and see what they do with it. Their "disposable income" was 158.4 billion dollars. This means all the money they could possibly spend. Technically, consumers' income was larger than this, since it included money paid out for taxes to the government and other obligatory contributions. But these are here accounted for under other headings. Of consumers' disposable income, they spent 143.7 billion dollars and saved 14.8 billion dollars.

What consumers spend is received largely by business (including, of course, agriculture and professional men). Since we have accounted already for consumers' spending, it would be duplication to set it down again at the gate of business income. Only that part of income which business does not pay back again, directly or indirectly to individuals, must be listed as business receipts. These undistributed profits and reserves amounted to 13.3 billion dollars.

What does business spend, outside of the payments for current services, dividends, and so forth? The remaining expenditure is for what is ordinarily called "capital" goods. New construction accounted for 8.5 billion dollars. This includes factories, railroads, commercial and industrial construction of all types. It also includes residential buildings. Most residential construction projects are undertaken by business concerns; those houses built by people who expect to occupy them are lumped with the rest for convenience. Producers' durable equipment— machinery and the like—took 12.4 billion. Finally, business may enlarge or diminish its stock of goods. In this case, the inventories were enlarged by 3.7 billion. This figure covers only goods actually added to inventories.

Revaluation of existing stocks of goods due to price changes is excluded, since that is not part of the money flow; it is merely a change in bookkeeping figures.

Altogether, business spent, for capital purposes, 24.6 billion dolars, or 11.3 billion more than it received as undistributed income. Of course, this does not mean that it operated at a loss. Business made tremendous profits in 1946. The deficit in this table merely reflects the fact that business invested other people's money, which is obtained by borrowing it or selling shares. This investment goes into the same column in which is listed the surplus which consumers did not spend, but saved. Part of the consumers' surplus was used to buy the mortgages, bonds, or stocks, by the sale of which business obtained the extra money to invest.

Part of the saved money flowed abroad rather than into spending or investment in this country. That part of the flow is measured at the gate called "International." Here we are not concerned with everything spent abroad, or everything contributed to our income by foreigners, but only with the difference between these two totals. We made available for the use of foreigners 4.8 billion dollars more than they made available for our use. Under normal circumstances this sum would constitute an increase of their debt to us; hence it is called an investment. Actually much of it was given away in 1946, but it must nevertheless be put down in the table, since the money must be accounted for.

Last of all is the gate of government. Here is put down *all* the money which was received by governmental units in this country and all the money they spent. This involves no duplication, since taxes and other payments to government by consumers have been omitted in listing their disposable income, and business taxes are deducted before undistributed profits are reckoned.

Government took in 56.5 billion dollars; it spent 55.2

billion. Thus there was a surplus of 1.3 billion to carry over into the last column.

What government spent went in part directly to consumers for wages and salaries. In part, it went to pay business for products and services; much of this sum business paid out again to consumers and the rest turned up in undistributed profits.

Now look at the right-hand column containing the excess or deficit of expenditure for each main item. It will be seen that they exactly balance out; the total of the column is zero. Individual consumers and government spent less than they received; this surplus or "saving" was used partly to increase domestic business investment, partly to increase foreign investment.

It should be noted by the way that not all the investment is included in the right-hand column. Before the excess of business expenditures over receipts was reckoned and carried across to the last column, business was recorded as spending 13.3 billion of its own accumulations. Taking account of this, the total number of dollars spent for investment was 29.4 billion, rather than 15.1. Exactly the same amount was saved. In such a tabulation, saving must always equal investment.

Using the Budget

At the beginning of 1947, when these figures were compiled, nearly everyone was concerned with curbing inflation—that is, halting the general rise of prices. How was this to be done? The main object was to prevent the total demand from growing. While some people had too little to spend, and many others would have been glad to enlarge their own incomes, a greater aggregate income would be almost sure to result in higher prices.

Here the amount of production enters the calculation.

Money is spent for goods and services. If the output of these could be increased, expenditures could be enlarged by the same amount without boosting prices. But there is a limit to the gain of production in a given time, and with a given amount of work. In 1946 there was no important reservoir of unemployed to call upon. To be sure, there had been big strikes, hence fewer labor disputes would mean more labor hours. Efficiency, too, might advance somewhat as new machinery was installed and people became accustomed to their new jobs. But a good estimate made by economists at the beginning of 1947 was that total output for that year might be about 5 per cent larger than in 1946—scarcely greater. No matter how much was spent, people could not buy any more goods than could be produced. It would be well, if possible, to keep total spending from growing at all, since in that case any gain in output could mean lower prices.

It is one of the elementary principles of economics, known to everybody, that when demand is greater than supply, prices tend to rise. In social accounting of the type represented by this budget, aggregate demand is total expenditure. These are just different names for the same thing. And one would not even have to know the elementary law of demand and supply to understand that if more money were spent for the same amount of goods, the price of the goods must be higher. That is an arithmetical necessity. If I pay $60 for two suits of clothes, the average price of the suits is $30. If in the following year, I pay $80 for two suits, their average price is $40. It is as simple as that.

Now, then, let us look at the table to see where the pressure for higher prices came from. A great many people said it came from high wages. Many other people said it came from large profits to business or farmers. Still others said it came from foreign demand for American goods. The first fact to notice is that it came from all these things put together. Every spending group contributed to the demand.

Consumer expenditures were obviously larger than those of any other group. But not all consumer income came from wages. There are also lumped in the figure the income from dividends, rents, farms and unincorporated businesses, and interest on private investments. Ignoring for the moment the taxes deducted to reckon "disposable" income, rental income was 6.9 billion dollars, dividends, 5.6 billion. If wages alone were kept from rising, some of these other items might grow enough to cancel the restrictions.

Nevertheless, one of the most important gates at which to control the flow of spending was that of consumers. One obvious way to keep this gate from opening wider was to retain price-control. For if consumers did not pay more for each unit, their spending could not increase, unless there should be more goods to buy. But suppose their incomes did increase. What would they do with the extra money? Clearly, they would have to save it. And, in that case, it would probably be invested at home (that is, spent for capital goods) or invested or given abroad, thus increasing the spending of foreigners. Even aside from black markets which may make price-control more or less ineffectual, incomes as well as consumer spending must be limited if total demand is to be limited. Therefore, it was important not only to have price-control, but to limit rises of wages and salaries and to control rents.

How about limiting business income? Well, if price-control could be enforced, business profits would be limited almost automatically. Business income as a whole was determined by two factors: (1) the number of things it had for sale; and (2) the prices at which they were sold. It could not get many more things to sell; therefore, if prices were prevented from rising, it could not make much larger profits—unless it paid less for labor, in which case labor income would shrink by a compensating amount. Freezing of prices of farm products would likewise keep farmers' profits from growing.

Price-control was, in fact, abandoned in 1946. But in 1947 business was urged not to raise its prices; labor was urged to go slow on wage increases. When some wage increases were made, business was advised to absorb the extra cost, either by greater efficiency which would release labor for more production, or by accepting smaller profits.

How about limiting foreign spending in this country? That would have helped a little to control inflation, but the net foreign demand was so small in relation to the total demand that it could not have helped much. And for other reasons it was not wise to cut this item of expenditure.

Finally, we come to government spending. This is a tricky item, and many people have misunderstood its effect. A popular slogan has been "cut government spending—here is where the pressure for inflation arises!" But remember that when government has a surplus, all the money it spends is raised by taxes or other contributions which individuals or business have to pay, and which consequently reduces *their* spending. Cutting taxes under these circumstances would merely give consumers and business more to spend. It would probably not reduce total spending by one cent.

If government reduced its spending *without* lowering taxes, that would help. The critical item here is the *difference* between government receipts and expenditures, that is, the cash surplus.

What happens to such a governmental surplus? It may be loaned abroad, in which case it increases demand. It may be used to pay off government bonds in the hands of individual citizens or business concerns. in which case they can invest or spend what they receive from the government. But it *may* be used to retire debt in the hands of the banks, in which case the money may be made to disappear into thin air. This is one of the mysteries of the banking system, which will have to be reserved for later explanation. To maintain a sizable government excess of receipts over expenditures is one of the best ways to fight inflation. The government thus takes money from the people which it does not give back; then it can wave a wand and presto! the money is gone.

What Happened in 1947

As most people will remember, inflation was decidedly not curbed in 1947 but took a new lease of life. This was particularly true in the latter six months of the year. The gate of price-control was not closed, and admonitions to keep prices down voluntarily were not very successful. There were wage increases, and the pressure of foreign demand also grew. Meanwhile production increased only about 5 per cent, as predicted. The annual rates at which the various important economic groups were receiving and spending money in the second half of 1947 as compared with 1946 were as follows:

Consumers' disposable income rose to 180.8 billion dollars from 158.4 billion in 1946.
Of this, consumers spent 169 billion dollars of 143.7 billion as in 1946.

Busines surplus income rose to 17.8 billion dollars from 13.3 billion.
Business investment was 31.7 billion dollars instead of 24.6 billion.

Net foreign investment was 8.2 billion dollars instead of 4.8 billion.

Government received from the public 58.7 billion dollars instead of 56.5 billion.
Government payments to the public *fell* to 52.7 billion dollars from 55.2 billion.

There is a particularly noteworthy fact about these changes. Government expenditures, alone among all the items, decreased. At the same time, government receipts increased slightly. Government was therefore the only factor which exerted an influence to reduce demand and hence to curb inflation. Here was a gate which shut a little way against the money flow. This result was made possible because taxes were not reduced. If they had been, consumers' disposable income and business income would have been correspondingly increased, and the reduction of government spending would probably have been without effect in limiting total demand.

It is also interesting that the difference between what consumers received and what they spent was narrowing. Their current savings shrank from 14.8 billion dollars in 1946 to an annual rate of 11.8 billion dollars in the last half of 1947. The increase of investment which occurred was financed not out of larger savings by individuals, but out of larger undistributed profits.

How far could this tendency go? Certainly the time might come when consumers, pressed hard by high prices, would cease to cut their savings and start to cut their purchases instead. Then business profits would quickly feel the effect, and business would invest less. With both consumers and business spending less, total demand would fall. If the fall continued, it could readily bring depression and unemployment unless government quickly reversed its policy and started to spend more than it received. Note that this conclusion is based not on a mere theory of "deficit spending," but on an inspection of actual figures of spending.

Comparison With Pre-War

Everybody knows that both national income and spending were much greater after the war than before it. But it is interesting to see actual figures for the main groups of spenders. These figures have been calculated so as to eliminate the effect of price increases. They indicate roughly the change in the actual quantities bought.

Consumers bought in 1947 about 48 per cent more than in 1939.

New business investment in 1947 was about 170 per cent more than in 1939.

There was an even larger percentage increase in "net foreign investment," but this item started at such a low level in 1939 that it did not make nearly as much difference as the other two.

Government before the war had a deficit; after the war it had a surplus. Therefore no percentage comparison is possible; but it was not making a net contribution to demand in 1947, as it did in 1939.

Whatever may be the nation's need for new capital, it can be said that in 1947 we were piling it up much more rapidly than before the war. The inflationary pressure resulting from expanded business investment was much sharper than that resulting from increased consumer buying. The same tendency continued in 1948.

Even if consumers did not start to buy less because of high prices, business might come to the end of its post-war investment program. Then business spending for new investment would shrink. This shrinkage might conceivably be just enough to stop inflation. But it it went beyond that, unemployment would grow, unless more spending from some other quarter (presumably government) counterbalanced the shrinkage.

The great value of the Nation's Economic Budget is that it makes possible estimates of such possibilities *before they happen*. Thus it is an instrument which can be used to forestall either inflation or unemployment if we understand it properly and will consent to take the appropriate action.

7

How Money Appears and Disappears

MONEY, we have seen, flows around and around, from consumers to business and government, from business and government back to consumers. (Money here is used in the broad sense to include not merely coins and "folding money" but also bank deposits, which are paid out by checks). We have also seen that for any given period, total expenditure must equal total income. If as much is always received as is spent (including investment), and as much is always spent as is received, how is it possible for income and spending ever to change? How does the income, which constitutes demand, ever get bigger or smaller? Experience tells us that it does.

The puzzle is something like that propounded centuries ago by the Greek philosopher Zeno about motion. A modern illustration of Zeno's puzzle might be as follows: Suppose you are equipped with a camera with a very fast lens and shutter. You take a picture of a runner doing the hundred-yard dash. This picture shows him at a given position on the track. No matter at what instant you take the picture, he will appear to be at one point or another. Even a movie film is a succession of pictures of the runner at a series of fixed points. How does he ever get from one point to the next? Where does the motion come in?

The answer is of course that for the sake of our "instantaneous pictures" we deliberately eliminate as much of the motion as we can in order to obtain a clear outline of our subject. If we left the shutter open, we should get a picture of the motion all right, but the runner would appear as a blur. The same is true of economic tables showing income and spending. Both income and spending actually change from day to day; we simply take a snapshot covering a single period at a time so that the picture may be clear.

In the case of money, however, we still have to account for the source of expansion or contraction when its volume changes.

Government as a Source of Money

Government can easily increase the amount of coins or paper money, and can do so without regard to the number of things produced. This can happen even if there is a rigid gold standard. On various occasions in the world's history new discoveries of gold, which could be sold by the miners to the government at a fixed money value per ounce, have increased the amount of money available and thus have led to booms. If a government departs from the gold or other metallic standard, it can pay its bills merely by running the printing press; this has happened many times.

Government can also reduce the amount of money in circulation by various devices, though this process is more difficult and less often resorted to. By heavy taxes it can build up a surplus and then destroy paper money collected or withhold money of any kind from further circulation. Or it can call in existing paper money, announcing that the old currency will cease to be legal tender as of a certain date. The money turned in is exchanged for new currency of lower denomination. For instance, the holder of a two-dollar bill might get for it a dollar of the new money. This is a method of enforcing devaluation, which has been practiced mainly after wars.

So-called budgetary inflation can occur when the government does not collect enough revenue from the public to pay its bills, and covers the deficit by expanding the money supply. It is a mistake, however, to assume that this practice will always lead to rapidly rising prices. When there is a large volume of unemployment and much idle industrial capacity, increase of money in circulation is likely to result in more production and employment. Only when full capacity for production is approached will prices rise very much, because only then will the means of payment grow more rapidly than production. During wars the labor supply is reduced by military service, and the demand for munitions is so large that production soon reaches the limit of capacity. Then, unless government collects from the public all it spends, as governments almost never do during wars, the money spent will be enlarged more rapidly than the goods and services produced.

Banks as a Source of Money

Government is not the only source of money in modern economies. Anybody who can borrow from a bank sometimes obtains purchasing power which neither he nor anybody else had before. If bank loans as a whole are extended faster than they are paid off, the supply of money (or deposits, which amount to the same thing) increases. When bank loans as a whole are paid off faster than they are extended, the supply of money inevitably decreases.

This may sound fallacious to anyone who has an elementary conception of the way banks operate. The ordinary supposition is that somebody with extra money deposits it in a bank; the bank then can lend this money to somebody else. How can loans be made if the money is not previously in existence? They can and are; anybody who wants to verify the fact can look at the tables in the *Federal Reserve Bulletin* and see how the total of bank loans changes from time to time.

When deposits grow, the reason is largely that borrowings have grown; the chain of causation ordinarily runs from loans to deposits rather than the other way. Somebody borrows from a bank; the bank adds to his deposit account the proceeds of the loan. If he draws it out by issuing checks, the amount by which his deposit is reduced appears as an addition to the bank accounts of those who receive the checks. The money loaned remains in circulation—either in deposits or in currency. When total borrowings from the banks are reduced, deposits (or currency) shrink by an equal amount. Nowadays, in the United States, many more payments are made by check than by currency. For instance, in January 1948 the total of bank deposits was about 145 billion dollars; the currency in circulation outside banks was about 26 billion—a ratio of nearly 6 to 1.

Banks do not expect all depositors to withdraw their money at the same time. That is why they can lend more than they have in cash. Experience shows, to take a simple illustration, that if a bank keeps on hand $10,000 in cash, it can under ordinary circumstances lend as much as $100,000 without ever having to pay out any of its reserve cash. While some people draw out cash, others deposit it. The reserve acts as a margin of safety. The bulk of withdrawals and deposits consist not of cash but of mere entries on the books. But if the bank's reserve should fall, say, to $7500, then it would probably have to reduce its loans to $75,000. An upper limit of loans in relation to the amount of reserves is enforced by law, but even if it were not, good banking practice would recognize it because of past experience.

When production of goods is growing and more people are being employed, business is likely to need to borrow more from the banks. But there is very little assurance that when bank loans expand, and in consequence the

total of demand grows, the production of goods and
services will go on increasing by the same amount. When
the country is producing all it can, with its existing
supply of materials, machinery, and labor, loans may
still increase. The mere fact that businesses need more
money with which to buy goods and pay wages does not
always mean that production and employment are increasing correspondingly. Sometimes the need for additional bank credit merely reflects rising prices.

In an inflationary period the interaction of prices and
bank credit is much like the behavior of a new military
missile called the "flying stovepipe." The flying stovepipe runs by a jet of hot gas expelled from its stern.
The gas is produced by a combination of compressed air
and burning kerosene or similar fuel. The missile takes
in air through its open snout; its speed compresses the
air; the more highly compressed the air, the more powerful is the jet. Thus (within limits of course) it is said of
the flying stovepipe that the faster it goes, the faster it
goes. Similarly, unchecked expansion of bank credit may
boost prices; higher prices in turn increase the need for
bank credit.

Nor is there any assurance that when the total of
bank loans decreases they will fall only to the point
where there is still enough demand to keep everybody
employed at capacity production. You may have a factory capable of making a hundred washing machines a
day and employing five thousand persons to do it. But
your sales may fall to eighty a day and your employment
to four thousand. If you had larger loans from the bank,
you could make more, but if you are a prudent manager, you will not borrow in order to make more than
you think you can sell, and if you are not a prudent
manager, the bank probably will not lend you the money
anyway.

Central Banks and the Money Supply

In all modern nations there are institutions which act as banks for the banks which lend money to the public. This fact makes possible a certain amount of unified control over the expansion or contraction of the supply of money.

In the United States the central banks are called Federal Reserve Banks; there are twelve of them, each covering a certain district. All are under the jurisdiction of a single Board of Governors. In most other countries there is only a single central bank, such as the Bank of England or the Bank of France.

An American bank which belongs to the Federal Reserve System (all national banks must belong to it, and many state banks belong as well) must deposit a stated part of its reserve with the Federal Reserve Bank of its district, keeping such cash as it needs for day-to-day operations. It can then borrow from the Federal Reserve Bank in order to lend more to its own customers. The maximum amount it can borrow is of course limited by the size of its reserve.

The Federal Reserve Board is granted by the government the power to issue paper money. Look at the bills in your pocketbook and you will find that, with the exception of those for one, two, or five dollars, most of them are labeled "Federal Reserve Notes." When a bank borrows paper money from a Federal Reserve Bank, this is the kind of money it gets.

A bank cannot borrow from a Reserve Bank just by asking for the loan. It has to put up collateral. Of course, when it gets any gold, it can obtain money or credit for that. A note signed by a businessman or a farmer who has borrowed is good collateral. Thus, if the Federal Reserve consents to lend it enough, a bank can keep on lending without depleting its money supply at all. It simply endorses notes in its possession, sends them in to the Reserve Bank, and receives from the Reserve Bank enough to lend just as much again. It must pay interest for these loans—usually at a rate somewhat below the rate it charges the customer. The interest rate charged by the Reserve Bank is called the "discount rate." By raising the discount rate, the Reserve Bank can do something to discourage borrowing; by lowering it, something to encourage the expansion of credit. Change in the discount rate is one of the powers of Reserve Banks to influence the volume of money.

Far more effective than changing the discount rate is the Reserve Bank's purchase or sale of securities. Suppose a Reserve Bank buys a thousand-dollar government bond. (It is not allowed to buy private bonds.) If it buys the bond from a bank, the payment immediately gives the bank $1000 more, either in cash or on deposit with the Reserve Bank. The selling bank's reserve is thus increased by the same amount. It can lend more money accordingly. If the Reserve Bank buys the bond from some person or company, the seller probably deposits the proceeds in a bank, and the same expansion of reserves occurs. The initiative in buying or selling bonds lies with the Reserve Banks; the initiative in discounting lies with the member banks.

When the Reserve Bank sells bonds, the checks (or currency) paid by the buyers reduce bank reserves, and the power of banks to lend is correspondingly reduced. Changes in the size of bank reserves may change the amount of loans which the banks can make by as much as six times.

This power of Reserve Banks to make money appear or disappear is not unlimited. Their loans and the currency in circulation in combination cannot exceed four times their gold reserve. They can increase the reserves of the member banks by buying bonds, but if the member banks do not lend a larger amount to the public, the money supply is not increased. Banks often do not lend all they legally can, especially in depressions. Those business concerns which need the money or think they can make profitable use of it may not enjoy good credit standing at such times. In time of prosperity and active demand for loans, the Reserve Banks can usually restrict lending by selling bonds. But even then, if a lot of gold is coming into the country to pay for exports, the banks which receive it have correspondingly increased reserves, and thus may counteract the effect of restriction caused by any Reserve Bank selling. (They must hand the gold over to the Reserve Banks, but of course they are credited with it.)

Government Power to Increase Bank Credit

Government *can* increase the supply of money by printing it, but most modern governments don't act so crudely any more. The government, during a war or a depression, has to spend more money than it is collecting. It borrows the money, giving the lender a note or bond. On the surface, this looks like an ordinary business transaction; it does not arouse the alarm which would result if the government simply printed a lot of money to pay its bills. If the lender is an individual or a company, and if he puts the bond away in his strong box and takes the money out of his bank account to pay for it, it *is* an ordinary business transaction and does not enlarge the amount of purchasing power. The buyer's deposit is decreased; the government's deposit is increased by the same amount; total deposits remain the same.

But suppose the purchaser of the bond goes to the bank and borrows enough to buy it. Then his bank deposit remains intact; the government's deposit is increased by the amount he pays for the bond, and the volume of money jumps a dollar for every dollar the government borrows. If a bank "buys" the bond, the bank lends directly to the government instead of to some bond buyer. Bank deposits are increased in either case, and the immediate effect is exactly the same as if the government had printed the money it spent instead of borrowing it.

Central banks almost invariably buy government bonds in periods of emergency. (Central banks are either directly or indirectly under government control.) By buying bonds the central banks give the other banks all the reserves they need to extend credit. Central bank buying also helps to keep up the price of the bonds so that the government will not have to offer a high rate of interest.

Thus government and the banking system in combination constitute a powerful means of increasing the supply of money. Government spends more than it receives; the bank loans to which its borrowing immediately gives rise enlarge the general purchasing power; the purchases of government bonds by central banks enlarge bank reserves and permit the banks to increase the means of payment still more by means of lending to private borrowers.

Between December 1941 and December 1947 marketable U. S. government obligations grew by about 120 billion dollars. Of these, commercial banks held at the latter date about 60 billion and Federal Reserve Banks 20 billion. This accounts for the increase in purchasing power which occurred. Deposits and currency in circulation together grew by about 90 billion in the same period. The total of deposits and circulating money at the beginning of 1948 was more than twice as great as seven years earlier. This was the primary source of inflation.

During the depression of the 1930's the government pursued the same financial policy as it did in the war, though its borrowing and spending were on a far smaller scale. There was some increase of money as a result, but not enough to eliminate unemployment. And because production did not grow to the level of capacity or even near it, prices did not rise markedly. For one reason or another, private borrowers did not make use of all the credit which the increased bank reserves would have made possible. Opinions differ as to what the reason was.

During the war bank credit did not grow so much as the increase in bank reserves would have permitted. This was largely the result of governmental controls of wages and prices. When these controls were removed in 1946, prices began to shoot up. Even at the beginning of 1948 further expansion of money and deposits was possible, on the basis of the existing bank reserves. And the Reserve Banks could increase these reserves still further by continuing to purchase government bonds.

Government Power to Decrease Bank Credit

When the government has added to the money supply by printing, it can reduce the supply again by collecting in taxes more than it spends. The same principle applies when the government has increased money by the modern process of borrowing. In this case, when it has a surplus of receipts over expenditures, it may reduce the supply of money by paying off some of its loans.

But the second process is more complicated. When the government redeems a bond from an individual owner the government's funds are reduced, but the seller's are increased by the same amount. The volume of bank deposits plus money in circulation therefore remains the same. In order to force a reduction of purchasing power in this way, the government has to do something to reduce the banks' desire or power to lend.

Again it comes back to what the central bank does. The central bank must *sell* government securities as the government buys them if the money paid by the government is automatically to disappear. In that case, the proceeds of the sales (deposits or money) find their resting place in the central bank, which does not put them into circulation again. As long as the central bank is selling, it is reducing the reserves of the banks which lend to private borrowers, so that they are likely to lend less.

Since governments nowadays control central bank policy, this result is comparatively easy to bring about. But the government must as a rule pay a price for it. When the government was selling the bonds, it maintained the price by central bank buying. Thus it obtained a low rate of interest. But now, if the central bank is to sell instead of buying, the price of bonds may fall and the rate of interest will rise correspondingly. This means that in future the government will have to pay more interest on a given amount of debt. This might be a serious burden on the budget and hence on the taxpayers. Banks are also faced with a problem because of the fact that a large part of their assets consists of bonds, which are falling in value.

For this and other reasons governments seldom deflate the money supply as much as they have inflated it. Serious deflation sometimes happens without anyone planning it, but when governments exercise deliberate policy in the matter, they usually are content to stop inflation, if they can, by making sure that no *more* money is added, instead of withdrawing large amounts from circulation and bringing about drastic cuts in prices. Serious deflation of this kind is adopted, as a rule, only when inflation has proceeded to extremes and an entirely new currency has to be substituted for the old.

Banks Are the Storage Reservoirs

Thus, in a modern economy, the banking system is the ultimate origin of extra money added to the flow of purchasing power, and the ultimate recipient of money withdrawn from it. Banks are somewhat like the storage reservoirs in a river system. When more water is needed, the gates may be opened; when less is wanted, the gates may be shut. Thus something can be done to prevent floods or droughts.

Government can take the initiative in opening or shutting the banking gates, and usually does so in wars or other great emergencies. It can increase the flow by borrowing and spending; it can decrease the flow by paying off its loans.

But in a system in which government is not the only source of spending—that is, in a system in which private enterprise plays an important role—government action does not suffice to regulate the flow in time of peace. What consumers and business concerns borrow and spend determines a large part of the flow of money and income. The banking system can do something to regulate this flow, but, on the whole, the banks are more passive than active in control of the flow of credit and money. In time of depression they cannot lend money which solvent borrowers do not want to borrow or cannot offer good security for. In time of great business activity and rising prices they have difficulty in preventing an increase of deposits and loans, unless government and central banks take a firm initiative.

There still remains the question why at some times the private sector of the economy borrows less than enough to keep the machines busy and employment high, and why at other times it borrows more than enough, and so produces speculative rises in prices.

8

Private Enterprise in Fancy and in Fact

WHY DO people sometimes increase the stream of spending and sometimes decrease it? What causes the irregularities of demand? This is the question with which economics has wrestled for years without making much scientific progress. It is the part of the subject in which the most uncertainty still prevails.

Classical economic theory has a back-hand answer, but the answer does not help much. According to this theory, there is not "naturally" any possibility of a shortage of demand. The money which is paid out in the process of production and sale must equal the amount necessary to buy the product. (This is Say's Law.) When troubles arise, they are merely because of temporary obstructions to the normal operations of competition and to the balancing of supply and demand of particular kinds of goods through the mechanism of prices. The adherent of this theory simply says, "Get rid of monopoly and other artificial regulations or interferences, and all will be well." More simply said than done! Presently we shall look at some of the obstructions to competition and price flexibility which exist.

The Keynes Revolution in Theory

First, let us examine a point of view which questions this major conclusion of classical theory itself. John Maynard Keynes, himself a professor and an adept at the intricacies of this theory, produced a revolution in the academic world by proving to the satisfaction of many that a level of full employment is not the "natural" state of things at all.

Whether or not Keynes was right, his ideas have created such a stir that it is impossible to ignore them in a discussion of the flaws of private enterprise. Just as a matter of general information, it is well to know what he thought. The actual formulation of the Keynes theory involves a series of mathematical equations and technical terms of the kind which characterize the classical theory itself; these are forbidding to the non-professional reader and consequently general knowledge of his view has been obtained second-hand and has been clouded by a good deal of misunderstanding. Nevertheless, it is possible to convey a rough idea of what he was driving at.

We have seen that total income always equals total spending plus investment, but that nevertheless income does change from time to time. Keynes argued that these changes take place largely through differences between the amount invested and the amount people refrain from spending with the intention of saving it.

Let us take an oversimplified example to illustrate the point. Suppose national income is $100. Suppose that of this amount consumers spend $90 for consumer goods, while the other $10 is laid aside by individuals and business in combination. Now, suppose business invests in inventory, machinery, plant, and houses, not only the $10 saved but $5 in addition. (It can do so by borrowing from the banks.) Thus, spending plus investment becomes $105. This, in turn, equals the national income. The national income has grown from $100 to $105.

Or suppose, on the contrary, that while $10 of a $100 national income was being withheld from the purchase of consumer goods, business invested only $5. Only $95 would thus be spent and invested, and the national income would have shrunk to $95.

The crucial point of the Keynes theory, which differentiates it from the classical view, is that equilibrium between total demand and total supply may be reached at a point of demand which is too low to maintain full employment and capacity production. The classical theory argued that the economy was always approximating equilibrium at a full employment level. The only things which could prevent it from remaining at this level were "frictions" and "rigidities" in the competitive price mechanism. Keynes held that, frictions or no frictions, equilibrium might be reached at a level which permitted unemployment.

The practical consequences of this conclusion are diametrically opposed to those which flow from the classical one. If you assume that the natural state of equilibrium is one of full employment, and departures from it are due to interferences with competitive private enterprise, then you must conclude that intervention by government is just as harmful as any other kind of intervention. In that case, government ought to refrain from spending to maintain employment, from fixing minimum wages, or from other forms of regulation. If, on the other hand, you believe, as Keynes did, that equilibrium may be reached at a level of activity where unemployment is prevalent, you would argue, as Keynes did, that government ought to make up for any deficiency of private investment by spending funds which it did not take from consumers.

A good deal of confusion has been caused in discussions of Keynes' theories by his assertion that total saving necessarily equals investment. This seems to contradict his belief that there can be a discrepancy between the amount actually invested and the amount withheld from current expenditure for consumer goods. The confusion arises from a failure to understand the exact sense in which Keynes used the words "investment" and "saving."

"Investment" as Keynes (like most economists) use the word does not necessarily mean what you or I woul mean if we bought a few shares of General Motors o A. T. & T. It means an increase in the total stock o "capital goods"—that is, an increase in buildings, ma chinery, business inventories, and the like. If you la aside money and use it to buy a share of stock which sell, and if I use the proceeds to buy an electric refrig erator, no investment in the Keynes sense has taken place The amount which you "saved" is ultimately spent, no for new capital, but for consumer goods. Your saving is matched by my dis-saving, so that the total of saving has not been increased. Thus, neither saving nor invest ment has grown. This is the sense in which "saving" equals "investment." The community as a whole can save only by increasing its stock of real capital.

It can happen, however, that business spends for new buildings, equipment, and stocks of goods less money than people as a whole divert from their purchase of consumer goods with the intention of saving it. If this happens, there will be a slump of employment, because there will be shrinkage in spending plus investment. Incomes will fall, people will be compelled to spend for living expenses what they have "saved," or to save less, or both. In this case, the amount of money being with-held from current spending will soon shrink to the amount being invested. The opposite process takes place if business spends more for capital than people are with-holding from expenditure. Employment will rise, in-comes will increase, and more money will be withheld from consumption until the amount "saved" again equals what is being invested.

Saving in the real or social sense is identical with investment. Saving in the colloquial sense of what people refrain from spending with the intention of saving it may for a time be larger or smaller than investment, but, according to Keynes, it will follow the trend of investment and in the end will equal it. At that point "equilibrium" occurs.[1]

In the United States the Keynes theory as developed by Alvin H. Hansen and others adds the doctrine of "economic stagnation." This idea grows out of the belief that it is theoretically possible for equilibrium to be reached at a point below full employment. The doctrine is that in this country equilibrium actually was reached at this point in the 1930's, because there was not sufficient lure for the investment of private capital. We were no longer a young country, with new territory to be opened up, rapid growth of population, and great new inventions to be exploited. Our economy had become "mature." While plenty remained to be done to bring about a higher level of living and more consumption on the part of the masses, this task did not offer sufficient return to private enterprise to call forth the necessary investment. More or less continuous governmental investment on a larger scale would therefore be required if full employment were to be reached and sustained.

Criticisms of the Keynes Theory

It is important to understand that while Keynes upset the classical notion of a natural equilibrium at full employment, he was trained in the classical way of thinking and phrased his theory in terms of an equilibrium of his own. In developing this theory, he dealt almost as little with the real world of monopolistic and other restrictions of the markets as did the older theorists whom he was attacking. Both he and they were talking not so

[1] In the figures of national income accounting, saving always equals investment. This balance is not the "equilibrium" Keynes is talking about; it is merely a mathematical consequence of the fact that for any given period total income must equal total expenditure. The figures would show a lack of equilibrium between what people withhold from spending and what is actually invested only by a change in investment and total income between one period and the next.

much about what really does happen, as revealed by measurement of economic statistics, but about what logically must happen, given certain assumptions or premises, which in many cases do not correspond with the actual world.

In this sense the Keynes theory is a hypothesis which needs testing rather than a final truth. It may be that the economic order does not naturally seek equilibrium at any point, high or low. It may be that the fluctuations of investment do not play the decisive role Keynes assigned to them. It is true that, according to the statistics, investment does fluctuate more widely than purchases of consumer goods in general. But it is also true that the outlay for consumer goods makes up a much larger part of the total flow of income than does investment, so that a small percentage variation in this part can be a much larger factor in changes of total income than the larger percentage variation in investment.

Nor is it necessary to assume that changes in consumer spending spring largely from changes in consumer income, as Keynes appears to do. When people spend so large a part of their incomes as in recent years for durable goods like automobiles, their demand for new cars may be financed by an expansion of consumer borrowing, or it may approach its limit and begin to fall off without any prior decline in their aggregate income. It is quite possible that consumer durable goods may play a role in fluctuations of demand that Keynes assigned to business investment. They are, in a sense, an important form of "savings" and "investment" themselves.

Before leaving the Keynes theory, we should perhaps learn to recognize one or two of the technical terms which he introduced, since they frequently recur in the discussions of economists:

"Propensity to consume" means the habit of spending a certain part of one's income for consumer goods. Keynes assumed that this was a fairly stable habit; as a conse-

quence, what was left, or saved, would tend to vary with
the size of the income. The existence of such a habit has
not been established beyond question.

"Liquidity preference" means the preference people
with savings may sometimes have to keep their money
in cash rather than invest it. This, when it comes into
play, naturally decreases investment. (Keynes held that
society as a whole could not hoard money; if money was
not spent or invested, it would, as a whole, disappear.
For the cash one person failed to spend would diminish
by that much some other income.)

Another important term is the "multiplier." This
means the amount which a given addition to spending,
or subtraction from it, will increase or decrease the total
income as it flows about the system. The total change
in income is likely to be larger than the change in spending which set it in motion, since money spent by one
person is again spent by its recipient and so on. The
effect wears off before the process builds up very far.
The multiplier itself differs according to circumstances.
Deficit spending by government during the New Deal
had much less effect in increasing national income than
Keynes expected; the multiplier turned out to be extremely small.

Concentration and Competition

Even if Keynes were wholly wrong, and the classical
economists were right, that a competitive system would
reach equilibrium at full employment if left alone, our
system has so many approaches to monopoly and so many
"administered" prices that the classical theory is almost
as irrelevant as if it concerned another planet.

Business combinations and the control of prices and
output through pools and gentlemen's agreements flourished after the Civil War. By 1904 the so-called trusts
controlled 40 per cent of the manufacturing capital in

the United States. Virtual monopoly control existed in thirty products, from asphalt to whisky, including electrical equipment, leather, petroleum products, glass, shoe machinery, steel, sugar, and tobacco. Anti-trust legislation and financial crises retarded, but did not destroy, the tendency toward concentration. It is estimated by Gardiner C. Means that in 1909 the two hundred largest non-financial corporations owned one-third of the business assets of the country (excluding banks, insurance companies, and other financial agencies.)

During World War I big business grew rapidly because war contracts went mainly to the great concerns, and these concerns made huge profits, a large portion of which they laid aside for future expansion. Their reserves enabled most of them to weather the storm of 1921. The giant business concerns increased in importance during the 1920's, and mergers continued.

The two hundred largest non-financial corporations, which controlled one-third of the business assets in 1909, had 48 per cent in 1929. Their percentage of the total corporate income grew from 33 in 1920 to 43 in 1929. Five per cent of corporations earned 78.9 per cent of the corporate income of the country in 1918, and increased their share until it reached 87.9 per cent in 1932. Those manufacturing corporations, each of which had an annual net income of 5 millions dollars or more, got 34.2 per cent of corporate manufacturing income in 1918 and 46.1 per cent in 1929. Those companies with annual net incomes of less than $250,000 lost ground, receiving 23.4 per cent of the total in manufacturing in 1918 and 19.1 per cent in 1929.

Big Business Grew Under Roosevelt

During the Great Depression all business suffered, and big business especially lost ground, because it exerts more control in the industries making capital goods, which

decline more when business is bad. During the recovery,
however, the tendency toward concentration made up for
lost time. The three hundred and sixteen largest manu-
facturers held 35 per cent of the working capital in 1926
and 47 per cent in 1938. Manufacturing corporations,
each having more than 50 million dollars in assets, owned
37 per cent of the total manufacturing assets in 1934 and
49 per cent in 1942. There were two hundred and five
of these giants in the latter year. In 1942 the manufacturing corporations receiving annual net incomes of 5
million or more had just over half the total manufacturing income as compared with 46 per cent in 1929.
On the other hand, manufacturing corporations with incomes of less than $250,000, which had received 19.1 per
cent of the total income in 1929, had only 11.6 per cent
in 1942.

In industries accounting for one-third the value of all
American manufactured products, the four largest producers in each industry turned out more than three-
fourths of the product of that industry.

The growth of big business under Roosevelt was not
primarily due to combinations or mergers of separate
companies but rather to expansion of individual concerns. Mergers, however, had thrived between 1919 and
1929. They were particularly numerous in iron and steel,
machinery, producers of food, motion pictures, retail
trade, and hotels. The concentration of holding-company
control in electric power was notorious.

There were nearly three times as many mergers in
1929 as in 1919. During the depression and the subsequent New Deal recovery, the number of business con-
solidations fell off sharply. Concentration then became
a matter of the growth of existing large companies.

Big Business Grew During the War

The two hundred and fifty largest manufacturing corporations owned 65 per cent of the nation's production facilities in 1939. During the war they operated 79 per cent of all the new facilities built with federal funds and privately operated. They held options to purchase everything they operated.

About 26 billion dollars was spent during the war for new productive equipment—plants and machinery. This compares with total manufacturing facilities existing in 1939, which had cost 40 billion. The new facilities are of the most modern design and the greatest efficiency. It is estimated that about 20 billion dollars' worth of the new capacity is usable for peacetime products. Nearly three-quarters of the outlay was to enlarge manufacture of the same product which the operator made before the war. Of more than a hundred thousand machine tools provided, 78 per cent were general-purpose tools.

Technological improvements and know-how, which make possible new types of products and cut costs of production, were developed during the war principally by the big companies. Research, much of it financed by the government, was conducted mainly in their laboratories. Out of war earnings, they spent millions to advertise their names and trademarks. After the end of the war the net working capital of the sixty-three largest manufacturing corporations was larger than that of all manufacturing corporations listed with the Securities and Exchange Commission in 1939, or about 10 billion dollars.

Of the 175 billion dollars of prime government contracts awarded between June 1940 and October 1, 1944, nearly two-thirds went to the one hundred corporations which received the largest orders. More than half of the plums were awarded to only thirty-three companies, each of which had orders totaling a billion dollars or

more. The first ten corporations received 30 per cent of
the contract awards (by value).

At the top stands the General Motors Corporation
(Du Pont controlled), with 13,812 million dollars of
war orders. Other motor and plane companies are near
the top of the list, which as a whole includes great cor
porations in many industries.

Did the big concerns share this business with the
smaller ones by handing out subcontracts? Only slightly
over one-third of the value of the prime contracts was
passed on to subcontractors. Of this amount, approxi-
mately 75 per cent went to other large companies. These
in turn made some further subcontracts. In the end,
however, according to an estimate by the Smaller War
Plants Corporation, the war production of small con
cerns amounted to only 30 per cent of the total.

The fact that a high degree of industrial concentration exists does not mean that there are not still many
small businesses. In manufacturing, 98.9 per cent of the
firms had in 1939 less than 500 employees each. These
firms employed 51.7 per cent of the factory workers. The
effect of the war on small business is shown clearly by
the fact that in 1944 the firms employing less than 500
apiece, which still made up 97.8 per cent of all manu-
facturers, employed only 38.1 per cent of the factory
workers. This shrinkage was particularly marked in the
war industries. In the non-war industries less change was
noticed.

Those small businesses which succeeded in obtaining
subcontracts became dependent on the large concerns
for their sales and materials. This dependency may continue.

The big companies have been using their huge liquid
assets to buy government facilities and smaller private
concerns. Since 1943 the rate of mergers and acquisitions
has been higher than in the previous fifteen years, and
has shot up almost perpendicularly. Iron and steel and

machinery accounted for 25 per cent of the total acquisitions between 1940 and 1945. Food and liquor, drugs
and pharmaceuticals, also have become more highly concentrated.

Prices Are Administered

One must not conclude that the big concerns never
compete with one another, or that all prices are set by
monopolies. There obviously is competition in the automobile industry, for instance. But the competition is of
a different sort from that which the classical economists
had in mind. The big producer has the power to set his
price and maintain it for a considerable period. He
knows what the others in his field are likely to do; the
interaction of this knowledge usually prevents much
price-slashing. What is more likely to happen, if high
prices reduce demand, is a cut in production and employment.

On the other hand, it will not do to underestimate
the extent of monopoly price-fixing. In a study of cartels
for the Twentieth Century Fund—"Cartels in Action,"
by George W. Stocking and Myron W. Atkins—the authors estimated that of 1939 sales in the United States,
87 per cent, by value, of mineral products, 60 per cent
of agricultural products, and 42 per cent of manufactured products were cartelized.

Concentrated control is not, of course, all on the side
of business. Labor organization has gained immense
power to regulate wages and hours through collective
bargaining. This may be as just as it is inevitable under
modern conditions, but it certainly does not lead to
flexibility of wages or to competition in the labor market. Minimum wages enforced by government cover un-
organized trades. Farmers, who normally would illustrate
the classical effect of competition on prices, are now
safeguarded by price supports and governmental pay-

ments of numerous kinds, and some system of the sort is likely to continue. Government agencies directly or indirectly control prices for transportation, power, and other services.

In these circumstances nobody can ever prove whether the classical economists were right in alleging that everything would automatically work out for the best under laissez-faire. We don't have it; we haven't had it at least for a century; we are not going to have it in the visible future. The hard facts make the argument irrelevant. It is, of course, still possible that the prevalence of inflexibilities of prices and costs does lead to the changes in total demand which we know exist. But in that case the remedy is not to recommend a non-existent and impossible non-interference. We must look for improvement in another direction. It must include planning and policy of some sort; it must hinge on decisions about prices and production made by those who have power to make them. The right decisions, we may assume as a working hypothesis, could lead to the equilibrium on a high level which traditional theory supposed would automatically result without planning. The task for social scientists is to discover what those right decisions are and how people can be induced to make them.

9

Planning for Employment

IN THEIR efforts to analyze the effects of business concentration, economic theorists have developed a number of new ideas. The old theory was content with expounding what would happen under the competition which it took for granted was characteristic of private enterprise, and, alternatively, what would happen under monopoly, which was the opposite of competition. One set of "laws" was supposed to govern prices, demand, and production under competition, another set concerned the behavior of monopoly. Both sets were deduced from what it was believed a businessman would do to gain the highest possible profit by selling a particular kind of article in a particular market.

But neither competition nor monopoly was very carefully defined. In actual economic life, there are all sorts of situations which differ from "perfect competition" on the one side, and complete monopoly on the other. Starting from the end of competition, economists began to discuss the changes of theory that would have to be made in the case of "imperfect competition"—that is, a situation in which, though competitive forces are at work, they operate slowly or incompletely. And starting from the end of complete monopoly, they worked back to "duopoly"—where two producers dominate a market—or "oligopoly"—where a few producers do so. Recently they have turned their attention to the situation where there is a monopoly or partial monopoly of buyers, as distinguished from sellers. The terms invented to describe this situation are "monopsony," "duopsony," "oligopsony." Playing with these terms, the agile theorist can go on to chart what he supposes would happen when a monopoly confronts a monopsony, an oligopoly an oligopsony, and all the other possible permutations and combinations.

Administered Enterprise

All this is a fascinating game for those adept at it and may yield important conclusions. It is still, however, largely in the speculative stage which characterized the main body of classical theory before much statistical measurement became possible. Meanwhile, there is developing another approach to the subject which holds that the kind of economy we have does not work according to the theories of competition on the one hand or of monopoly on the other, and cannot be explained by a theory that there is a mixture of the two. A large part of industry is controlled by a type of organization which is a different sort of thing altogether, and is not seeking primarily to maximize profit in the sales of a specific article in a specific market.

The great business organization, it is pointed out, is not today controlled by "owners"—that is, stockholders. It is governed by managers or executives who, though technically appointed by a board of directors who in turn are legally elected by stockholders, are in fact pretty much on their own, once they have gained high position. These managers of course have a responsibility to keep revenue above costs over the long run; in that sense they are concerned with profits. But, in such an organization, profit is usually a sort of by-product which results from a host of decisions about policy of many different kinds.

It is a leading aim of the manager to safeguard the future stability and growth of the enterprise, not just to make the largest possible profit on a particular transaction. He exercises a good deal of control over the cost of what he buys. He can enlarge the "cost" of operation by adopting one kind of treatment of depreciation, inventory valuation, and reserves; he can reduce it by another kind of treatment of these accounts. Profits can be taken in one year or projected into the future according to the type of accounting used. Indeed, it is common to have simultaneously at least two different sets of accounts, one to satisfy the income-tax authorities and another to satisfy the Securities and Exchange Commission and to make public.

Prices in turn are set, not primarily by the competitive forces of the market, as the traditional economist understands it, but by decisions of the seller. This does not necessarily mean he sets the price to maximize profit on each given article. He may sell dozens of kinds of articles, some of them at no profit or a loss, others at a very high margin. He is governed by rules of thumb, custom, general policy for the future, considerations of prestige, desire to keep good will. He may control raw materials, semi-finished products, finished products; his operations may extend all the way to the consumer. He may take a profit at one point in this line and sacrifice it at another. Once a price structure is set, it has all sorts of interrelationships which are difficult to modify; nevertheless, it may have to be modified, both generally and in detail, because of technological changes, changes in "cost," changes in demand. The kind of price that is set by the seller in the hope that it will have some degree of permanency but nevertheless will not be completely under his control, since conditions may force him to modify it, is not primarily either a competitive price or a monopoly price or any simple cross between the two. It is an "administered price."

The modern industrial manager has to make administrative decisions about things other than prices which have an important effect on the economy as a whole. Under the stress of collective bargaining, he decides about wages, hours of work, and other elements of labor cost. He decides how much of the "earned surplus" shall be paid out in dividends. He decides what new investment shall be made and when, how much of its cost shall be paid out of the corporation's own savings and how much by borrowing or selling securities. He decides the rate at which production or purchasing shall go on—a rate which is not necessarily parallel to the amount of sales. He decides, in other words, when to build up inventories and when to cut them down. The "industrial manager" does not necessarily mean a single dictator of a company's fortunes. Decisions of this sort arise out of advice, careful discussion, or even out of a body of rules and practices worked out with the aid of research and experience.

A recent book on this type of business, *Managerial Enterprise,* is by Oswald W. Knauth, who has the advantage of having both the training of a professional economist and extensive experience in business management of the new type. Mr. Knauth estimates that managerial enterprise controls about half the production in the United States, the other half being divided between free enterprise of the old competitive kind and governmental or, as he calls it, collective enterprise.

In Mr. Knauth's opinion, the manager is in reality a sort of mediator in behalf of the company as an institution. He mediates among the contending forces of owners, employees, customers, and government. His decisions affect the fortunes of all of them, the future of the corporation itself, and, of course, the general welfare of the country and the world. But this type of calling is comparatively recent and is not governed by any clear set of standards. The manager of a great concern usually has, for the time being, a wide margin of discretion in which to make his important choices without wrecking his enterprise. He has lost the old and clear imperatives of competition, yet nothing very definite has been substituted for them. He usually has ideas about the proper kind of industrial or public policy, though there is little assurance that in any given case these ideas are the best ones.

Government's Pressure on Business

Government has long exerted many kinds of pressure on business, both direct and indirect. An old endeavor by government is to attempt to dissolve monopolies and force business to compete. Though successful in some respects, this policy has done little to prevent the growth of industrial concentration or managerial enterprise. Government has also had a finger in the regulation of rates and prices, particularly in railroads and public utilities. It has prescribed the methods and standards of labor policy. During emergencies its regulation has extended over a much wider field. Its taxation laws and rulings have had a profound effect on business decisions. These are only the more important types of governmental intervention.

Part of the lack of public standards for industrial management has arisen from the confused effect of governmental pressure taken as a whole. During war the confusion is lessened, though the pressure is greatly increased. War necessities put before the nation certain major objectives, such as increased production of munitions and food, the direction of resources and manpower into the most essential uses. Big business has had signal success in adjusting its activities to the national effort under these circumstances. Though in detail the record is spotty, the general achievement is attested by the remarkable increases of war production. But in peace managers have often been at a loss to know what they lawfully could do or not do in respect to competition, trade practices, collective bargaining, accounting, tax payment, and so on. Their usual effort has been, under ordinary circumstances, to try to avoid getting into trouble with the law while pursuing whatever course happened to seem advisable at the time, rather than to co-operate for positive purposes in a national program.

Among economists there are various schools of thought about what attitude toward business should be adopted by government. As far as business organization is concerned, there are three main doctrines. The first is the one traditional in this country, that government should force business to compete. Once it has done that, these economists believe, competition itself will see that business serves the public welfare. Regulation is of course permissible to soften the extreme rigors of competition where it is most severe—as in agriculture or the labor market. Regulation or public ownership is also required where monopoly is unavoidable in essential public services—as in railroads or utilities. Beyond this, however, the goal is "free, competitive private enterprise," enforced by government.

A small (in the United States) but persistent minority holds to the theory that private enterpise will not in most instances serve the public welfare and that the solution lies in the direction of socialism. Public or co-operative ownership is advocated at least for important or basic industries; some would go further than that.

The doctrines emphasized by Keynes may be combined with either of these attitudes toward business organization. It is possible to hold that the government can, and should, compensate for irregularities of employment by its own policies of spending and taxing, and at the same time to believe in letting businesses run themselves so long as they compete. It is equally possible to argue that the whole difficulty would be avoided if the government itself regularly made most of the decisions about business investment, prices, and production, as it would if it owned industry.

A third general point of view about business does not place its emphasis upon either governmental policing of business or governmental ownership. No matter what form business management has or may take, this school stresses something new—the choice of a peacetime goal, the working out of definite policies to approach that goal, and the co-operation of government, business, labor, and agriculture in the attempt to apply those policies. Implicit in this attitude is the assumption that business managers have much discretion to decide about prices, volume of investment, and the like, that it is possible to work out guides for these decisions which will accord with the public welfare, and that business managers, as well as labor and farm leaders, can in fact co-operate with the government in seeking the declared goal. At least it is desirable, according to this view, to work out coherent standards for decisions and to give industrial managers the opportunity to co-operate before trying more drastic methods.

The Employment Act and the Economic Council

In 1946 Congress passed the Employment Act, which set a leading economic objective for the country and established an official agency and procedures to seek the objective. The objective is the maintenance of a high level of employment and production. This law was inspired partly by those who wanted a governmental fiscal policy to compensate for the ups and downs of demand originating in the operations of private business, and partly by those who believed in trying to enlist the co-operation of business in stabilizing the economy.

The main objective is defined in the law as follows:

The Congress hereby declares that it is the continuing policy and responsibility of the Federal Government to use all practicable means consistent with its needs and obligations and other essential considerations of national policy, with the assistance and co-operation of industry, agriculture, labor, and State and local governments, to co-ordinate and utilize all its plans, functions, and resources for the purpose of creating and maintaining, in a manner calculated to foster and promote free competitive enterprise and the general welfare, conditions under which there will be afforded useful employment opportunities, including self employment, to those able, willing and seeking to work, and to promote maximum employment, production and purchasing power.

Under this law, the President must submit to Congress at the beginning of each session an Economic Report reviewing the existing economic situation and the trends, stating the level of employment, production, and purchasing power needed to carry out the declared policy, and containing a program for carrying out the policy, together with recommendations for any desirable legislation. Supplementary Economic Reports may be made during the year.

Congress must establish a joint committee from the House and Senate to review this Report. The committee is bound by law to state its position on the recommendations of the President and to sponsor bills carrying out any suggestions which it approves.

The Economic Reports are prepared with the advice of an expert Council of Economic Advisers consisting

of three men with a relatively small staff. This Council
is part of the Executive Office and is directly under the
President. It also consults with business, labor, agricul
ture and consumer organizations in the course of formu
lating and carrying out the policy of the law.

The enormous statistical resources of the government
are available to the Council. Its reports so far have con-
stituted the most comprehensive picture of the economic
situation ever made available to the officials and people
of the country, and have contained the first regular official economic programs aimed at achieving a leading
objective defined by law. The Nation's Economic Budget,
described in Chapter 6 of this book, was first presented
by the Council in its present form, though many other
economists helped in the development of the idea.

The nation is now equipped, therefore, with a series
of economic instruments which it never had before the
end of World War II. It has:

1. An enormous amount of current statistics which,
   when pieced together correctly, give a revealing picture of the existing situation.
2. A declared economic objective for peacetime, embodied in law by Congress: maximum employment,
   production, and purchasing power.
3. A legally authorized body of experts, in the office of
   the Chief Executive, to interpret the facts and tell
   us what ought to be done to seek the objective.
4. A major committee of Congress to study the Economic Report of the President and recommend
   legislation.
5. A legal authorization for the expert advisers of the
   President to enlist the co-operation of business, agriculture, labor, state and local governments in seeking this end.

Government Spending and Taxing

According to the Keynes theory, as well as according to the figures of the Nation's Economic Budget, it makes a great deal of difference to the state of the economy whether government spends more than it receives, or receives more than it spends. There is a substantial body of opinion in favor of so arranging taxes and government spending that government will have a surplus and will pay off part of its debt in time of prosperity when inflation threatens, and will, by spending more than it receives in time of unemployment, make up for the shrinkage in private spending plus investment.

It is implicit in the theory which gave rise to the Employment Act—though not stated in its words—that this policy should be followed. During its first two years of operation, which were years of full employment and inflationary pressure, the Council of Economic Advisers strongly recommended that taxes not be reduced; the President adopted this recommendation in his Economic Reports to Congress and reinforced it by vetoing tax reductions passed. The fact that Congress finally enacted tax reduction over the presidential veto is one of the indications that the recommended policy—called in economist's jargon a compensatory fiscal policy—is not always easy to effectuate.

Political pressures, as has often been pointed out, make it difficult for governments to adjust their policies of taxation and spending promptly enough—or sometimes even in the right direction—to compensate for too much or too little demand in the private sectors of the economy. Even if such political pressures did not exist, the job would be difficult. There is a real need for the products of government spending; budgets are not very flexible. If we ought to have more schools, roads, river improvements, or armaments, we cannot deliberately post-

pone them too long just because there is an excess of
demand for automobiles. During a depression the government can find plenty of useful things on which to
spend extra money, but even then it might be better
if in some way the money could be spent for food and
clothing rather than for dams and parks. It is easy to
produce a government deficit, when it is needed, by
reducing taxes; but it is not so easy to produce a needed
surplus by raising them. Something can be done by the
compensatory fiscal policy, but government is only one
factor in the national income. The Council of Economic
Advisers has turned its attention to other factors as well.

Price, Wage, and Investment Policy

Government, business, and labor together can, if they
will, exert a profound effect on the flow of income and
spending in the private sector of the economy. All carry
out policies which directly influence prices, wage incomes, and investment, in addition to the indirect effect
of government taxing and spending. But in order that
such policies may be co-ordinated about the national
goal, it is necessary in the first place that desirable measures be clearly known and stated. This is made possible,
for the first time in our history, by the use of such instruments of analysis as the Nation's Economic Budget and
the statistics of employment.

We know roughly how large the labor force of the
nation is, and population statistics enable us to predict
what it will be in the future, at least in the near future.
We know how large a national income that labor force
can now produce if fully employed. It is possible to make
fairly accurate guesses of probable increases in productivity, so that one may estimate how large an income
can be produced by the labor force which will exist two
or three years hence.

Armed with this information, the economists can project into the future the size of the national income which will correspond with a fully employed population, assuming that prices do not go up or down. If the money income should be larger than this, the necessary consequence would be higher prices—that is, more inflation. If it should be smaller, unemployment and less than full production must be the result, unless prices should fall correspondingly. Neither inflation nor unemployment is desirable.

The next question is, how is an income of the required size to be composed? If we look back at the Nation's Economic Budget for 1946 on page 66, we see how receipts were divided in that year among consumers, business, net foreign investment, and government.. We also see what expenditures were made by each of these great groups. But this is not the only possible division. It was altered in 1947 (all shares except that of government having risen) ; it will be different again in subsequent years. The part which each group takes of the total income cannot change too much in a short time; nevertheless, there is a certain elasticity which makes foresight and decisions on policy possible.

The United States now has the rudiments of a foreign trade and investment policy on account of the European Recovery Program and other measures of foreign aid. We calculate in advance what the net foreign investment is likely to be. We know what gross domestic investment has been—in the second half of 1947 it was running at the rate of 31.7 billion dollars a year. What it will be in the future depends on the amount of new construction, the purchases of producers' durable equipment, the changes in inventories. All these things may be roughly forecast; all are subject to policy decisions, directly or indirectly. Government expenditures and receipts are also under some control. We know, too, the existing purchases of consumers' goods and the trend of their production. We know how much consumers have been saving and can guess how much they are likely to save in the future.

Given these elements, it is possible to construct what economists call "models" of the national income for a given future year. There may be a number of different models for the same year, each based on different assumptions. You may assume a falling off of domestic investment. You may assume greatly expanded government expenditures or reduced taxation. Your assumptions, if they are to be of much use as guides, ought to correspond with something that could actually happen. At any rate, the purpose of any assortment of models is to indicate what the possible choices are. If, for instance, domestic investment is to be reduced, either other items of expenditure must be increased or the total income will fall. Or, to take another instance, if government spending is to be larger and tax receipts smaller, either there must be reductions in spending by consumers and business or prices must be higher.

The construction of a model for income of a future year is not a matter of pure fancy. It must correspond mathematically with the equation—income equals product. It must also take due account of the income flow. Suppose, for instance, you construct a model on the assumption that there is to be a drop in net foreign investment and in business investment at home, while government spends less and taxes more. These developments, you will have to conclude, will reduce the national income unless there can be a corresponding rise in the expenditure of consumers. But if foreigners, business, and government all cut their expenditures, consumers' incomes will probably be reduced, since these are the main sources of variation in the incomes of consumers. This model budget may show that in order to achieve a national income large enough to maintain employment at current prices, given the assumptions you have made, consumers' expenditures as a whole must be larger than their incomes. That is a mathematical impossibility. Indeed, consumers as a whole will probably save *something*, no matter how small their income is. The conclusion from your model therefore is: if foreigners, business, and government cut their spending as much as you anticipate, there must be either unemployment or lower prices.

By the use of model budgets for the nation's economy, the possible choices of policy may be indicated, and a combination of policies which could maintain maximum employment and production may be chosen. Once this were done, those who have power to make economic decisions would have criteria for action. Such power resides in all sectors of our economy. Some decisions may be made by government, some by business administrators, some by labor, and some even by the lowly consumer.

Economic science therefore has progressed to the point where it can make a realistic appraisal of possible choices for policy. Science *does not* prescribe which combination of policies must be chosen. It *does* show which combinations are internally consistent, and which of these are consistent with high employment.

Production Goals Versus Avoidance of Depression

During the depression of the 1930's the prevailing opinion among advanced economists was that private business must and will pursue its incalculable way, and that the most government could do would be to compensate for its vagaries. When war brought prosperity of its own kind, attention was shifted to avoiding the depression which experience had shown was likely to follow it. More recently, another attitude has been gaining ground. It might be possible, it even might be practical and effective, to substitute for these negative views a positive program. Instead of asking how a depression could be avoided or compensated for once it had arrived, we could lay out a program for producing whatever the nation most needed and then try to keep production high in gaining these goals. We know that needs here and in the rest of the world are far greater than the utmost possible total of production can satisfy. Depression would necessarily be avoided if only we could keep busy making and distributing urgently required goods and services. Not anti-depression measures, but pro-production measures, may yield the greatest benefit.

The Council of Economic Advisers, in drafting the Economic Reports of the President, has begun concretely to outline objectives of this kind and to indicate what measures are necessary to achieve them. In the Report for January 1948 it surveyed briefly the possibilities of increased income in the future and the means of enlarging production. It estimated conservatively that ten years hence there can be a per capita disposable income "about 80 per cent above the level of 1937 and 27 per cent above the level of 1947 in terms of constant dollars. . . . We have within our reach an economic environment that would make it unnecessary for masses of people to be undernourished or ill-housed, to work in obsolete plants or shops, or to lack essential medical care, social security, or education."

The report discussed our resources of land, water, forests, minerals, and what should be done to preserve and make more efficient use of them. It indicated what was required to increase industrial capacity so that there might be an efficient balance among industries. It outlined needed effort and expenditure in transportation, urban development, housing, education, health, old-age security. It then discussed more general policies adapted to achieving long-term goals.

This is merely a beginning. The pointing out of possibilities or needs, and the discussion of policies, is a long way from assuring that all concerned will take the required action. It is a great advance, however, that now for the first time all these subjects can be discussed, not as vague abstractions, but in the setting of fairly accurate knowledge of what it is physically possible to produce and what our income looks like. Economic science has at last begun to give us the mental tools by the use of which national economic policy can be something better than an expression of prejudice and something more substantial than a daydream.

The Great Unanswered Question

Knowledge, it has been said, is power. Power, however, has to be picked up and used if it is to achieve anything, and it must be used for good ends if it is not to be destructive. We still do not know whether the kind of society we have is capable of making good use of the new science of economics. Will those who must make decisions pay sufficient attention to the organized facts? Will they accept as criteria for action the standards which the new knowledge has begun to provide? We do not know.

Some have already made up their minds that the hope is vain. Private enterprise, perhaps even human nature itself, they are sure, is not equal to the test. Too many special interests, too many group pressures, are at work. These pessimists may be right, but it is necessary to point out that there is little scientific evidence for their gloomy conclusion, since the test has never been made. We cannot tell what people are capable of until they are provided with the knowledge and the standards which would

enable them to know what they ought to do. The rudiments of this knowledge are now available to Americans
for the first time in their history. The first experiment
in enlightened and voluntary economic government is
at hand. Democracy has already proved its compatibility
with great achievements in detailed management of industry and agriculture. Perhaps the fate of the world for
centuries to come depends now on the capacity of democracy to do a creditable job in management of the economy as a whole.

10

The International Income

THERE IS A habit of thinking which regards nations only as economic competitors in the sale of goods. Each nation, according to this idea, should strive to sell as much outside its borders as it can, and to buy as little. The greater the difference between its sales and its purchases, the greater its profit. The first systematic exponents of this view, which flourished in Europe before the American Revolution, were called mercantilists, because they regarded a nation as a competitive merchant.

Economists of every school from Adam Smith on have had no difficulty in demonstrating the absurdity of the mercantile thesis. If every nation were only to sell, who would do the buying? Where is the purchaser who can keep on buying indefinitely without earning anything? Furthermore, what is this "profit" which a nation gains when it exports more goods and services than it imports? Obviously foreigners are getting more real wealth from it than they are yielding in return. They may, for a while, pay in gold, but the gold itself is of no use unless it is to be spent. Gold cannot be spent by the nation receiving it unless at some time that nation buys more abroad than it sells.

The process of social accounting which is used to reckon the national income may also be applied on an international scale. There is a tendency for the incomes of all nations to rise or fall at the same time, just as there is for the incomes of all the groups within a nation.

This is a necessary consequence of the fact that the purchases of any nation constitute the sales of other nations. To have figures for the national incomes of the United States, Great Britain, Canada, France, and other countries, without any figures for the world income, is almost as fragmentary a way of dealing with the subject as it would be to have figures for the state incomes of New York, Pennsylvania, and Ohio without knowing much about the national income. There are already rough guesses of world income, and someday the figures are likely to be more precise. At present an adequate statistical basis is lacking, except for the more advanced countries.

One thing we may be sure of. Any world account of foreign trade must show that the sum of expenditures of all nations outside their borders exactly equals the sum of their receipts from outside. For every spender, there is a corresponding receiver, just as within a nation. The world cannot buy more than it sells, or sell more than it buys. Some nations may have a deficit in their trade, others a surplus, but the totals of the two groups balance out. (Accounting of foreign trade is far more complete and accurate than accounting of national incomes in backward nations, and it supports this self-evident truth.)

The American Giant

The figures for the national income of the United States account for a considerable slice of the world income. Approximate estimates indicate that this country, with about one-fifteenth of the population, land area, and natural resources of the globe, turns out more than half the world's industrial production, and probably about one-third of all the goods and services of every description. Even by 1929 the national income of the United States was equal to that of twenty-three other countries combined, including Great Britain, Germany, and France.

The consequences of these facts are momentous. The sheer economic bulk of the American Republic renders it the chief influence in stability or instability of the world's economy. We could not be greatly injured by any misfortune in, say, Argentina or Sweden, but other nations would necessarily be affected by any important change in our income.

Foreign nations not only require machinery, trucks, cotton, tobacco, petroleum, and a host of other products from us, they need our orders for what they produce. Ordinarily we do not think of the United States as an importing country. It is not so dependent on imports as most others, yet it is the largest single importer in the world next to Great Britain. In the years 1936–38 our imports were valued at nearly 2.5 billion dollars a year; in 1947 they were more than 4.6 billion dollars.

When employment shrinks in the United States, Americans eat less sugar and bananas, drink less coffee and tea. American industry buys less tin, nickel, chromium, manganese, pulp and paper products, wool, rubber, and dozens of other materials. South America and Asia are then thrown into even deeper poverty than usual; Australia, Indonesia, and Africa suffer. In consequence, they buy less from Great Britain and European countries; the manufacturing nations in turn suffer unemployment and cannot feed themselves. In spite of the fact that the bulk of British foreign sales (mainly manufactured products) do not come to the United States. British exports before the war followed almost exactly the same course as American industrial production.

The Tangle of Payments

Income equals product on the world scale, just as it does on the national one, yet there are bothersome differ-

ences which make the international economy more deli-
cate and intricate. There are no tariff duties or other important controls of trade between, say, Illinois and Pennsylvania. Both states use the same currency. Anyone in
Illinois who wants to travel in Pennsylvania, or decides
to move there, can do so if he has the necessary means.
Neither state restricts immigration from the other. The
result is that people in Illinois can buy and sell in Pennsylvania to the extent that cash or credit is available. If
one state should prosper less than the other, unemployed
workers could look for jobs across the state boundaries. A
man who has capital to invest does not worry because
state lines may intervene between his investments and his
place of residence.

But nations do interpose trade controls at their boundaries. They use different currency units; each important
nation has its own monetary and credit management.
They require passports and visas; they restrict immigration; they impose customs duties. In consequence, it is
often necessary to be concerned with the relative values
of currencies, the interrelationship between employment
and the exchange rate, the difficulty of paying debts across
national boundaries.

Economists have worked out a system of accounting for
the international payments of any given country. Such
an account is based on the axiom that the payments coming into a country in any given period must equal the
payments going out, if changes in indebtedness—called
loans or investments (that is, capital movements)—are
included. The same principle could be applied to the ex-
ternal trade of any state or section of the United States,
but ordinarily is not, because there is not so much need
to pay attention to it, and records of transactions over
state boundaries are usually not kept separately from
those of transactions within them.

If the reader will look back at the Nation's Economic
Budget on page 66, he will see under the heading "Inter-

national" an item for 4.8 billion dollars of foreign investment for the year 1946. This is the difference between
the value of all goods and services received by foreigners
during the year from the United States and all those that
were received from foreigners by Americans. In other
words, it is the year's addition to what foreigners owed
us (or were given). The item is a minus one in the budget
because we supplied this part of our product without
getting anything currently in return. This accords with
the fact that an export surplus is a subtraction from the
income of the country that sends abroad more than it receives, though on grounds of policy it may sometimes be
desirable.

A summarized statement of the "Rest of the World" account of the United States for the year 1946 shows how
this balance arose. This statement appears below.

Rest-of-World Account, 1946

(MILLIONS OF DOLLARS)

Net Payments of Factor Income to the United States:		Net Capital Movement from the United States:	
Wages and Salaries	8	Long term	3,342
Interest	122	Short term	1,176
Dividends	118	Change in gold stock	623*
Branch Profits	198		
Net Purchases from the United States			
		Errors and Omissions	—118
From Business	4,285	Adjustments for United States Territories	
From Government	1,139	and Possessions	—250
From Persons	—1,097		
Net Current Payments to the United States	4,773	Net Disinvestment in the United States	4,773

* When the United States *receives* gold, this entry has a plus sign,
as here.

From the items on the left-hand side of this account, it
may be seen that in almost every category, in 1946, foreigners paid more for our services and goods than we did
for theirs. (In order to get these "net" figures, outgoing

payments are subtracted from incoming ones.) Thus, they paid 8 million dollars more in salaries and wages to Americans than Americans paid to them. On balance, interest, dividends, and branch profits flowed in this direction. Of course foreigners bought huge amounts of goods here, partly from business, partly from government, while we bought much less from them, in part because in 1946 they did not have much for sale. Only the purchases of individual Americans abroad (chiefly tourists) exceeded foreigners' purchases from American individuals. The surplus of payments to Americans naturally is the value of the excess of exports from the United States over imports to it, including both trade and payments for services, or "invisible exports."

The right-hand side of the table shows how, in general, this excess of exports was financed. Long-term and short-term credits were both used; under these totals is also included the sale here of already existing foreign securities. Naturally the details would show that the capital movement was largely a matter of United States Government credits, such as the loan to Great Britain. In addition, foreigners sent over 623 million dollars in gold to help pay for what they needed.

Since not all international transactions can be traced, there is not a perfect balance, and consequently a relatively small item for errors and omissions has to be included. An adjustment also has to be made on account of the fact that while most of the figures cover continental United States only, some also cover the outlying territories and possessions.

Accounting statements like this are merely records of the truisms about foreign trade which economists have long been reiterating. For example, if the items on the right-hand side of the account had been smaller, foreigners could not have bought so much from us. They either had to borow the money or pay in gold. Why did they have to borrow? Only because we did not buy from them nearly as much as they bought from us. If the items at the left-hand side of the account had added up to zero, foreigners would not have had to borrow a cent or send us an ounce of gold. Current payments would have been in balance.

Monetary Fund, World Bank, ITO

For many years before World War I the foreign exchange values of most currencies remained stable, and while it was not so easy to buy or sell abroad as at home, people were not deterred by any fear that the relative values of pounds, dollars, francs, and the rest were likely to shift widely. The traditional economists had a ready explanation for the stability of exchange rates—an explanation which seemed self-evident to those who dealt in foreign exchange. Suppose a man with pounds sterling wanted to buy something in the United States. Naturally he would need dollars for the purpose. He would buy dollars with his pounds. If the demand for dollars should exceed the supply, the price of dollars (that is, the exchange rate) would rise. This would mean that it would take more pounds to buy in the United States. But the possible rise was limited by the fact that the United States and Great Britain were on the gold standard. Anyone could buy gold with his pounds, and buy dollars with his gold, at rates permanently fixed by the governments. So, instead of directly buying dollars after the rate had risen above a certain point, the British purchaser would buy gold and ship it to the United States. The limits within which the exchange could fluctuate were determined by the cost of shipping gold.

This was true so far as it went, but it did not reckon with the possibility that nations might abandon the gold standard, as many have since been forced to do. Obviously, if the demand for dollars was greatly in excess of the supply, and continued so for long, shipments of gold to the United States would be so large as to drain off the gold reserves of other nations. During World War I foreign purchases in the United States were enormous, and so gold shipments had to be stopped. The exchange rate was arbitrarily fixed by the governments concerned. After that war the world tried to return to the gold standard, but the attempt did not work for long. In the depression of the 1930's exchange troubles multiplied, the gold standard was abandoned, and World War II gave the final blow to any hope of restoration of the gold standard by Great Britain and many other nations.

In planning for the restoration of world trade, which it was hoped would follow the recent conflict, one of the first things that had to be done was to make some systematic provision for the stabilization of exchange rates, to act as a substitute for the gold standard. The International Monetary Fund was set up for this purpose. Nations which join it are obliged to contribute capital according to quotas based on their economic importance. The contributions are partly in gold, partly in their own currencies. Exchange rates are fixed by the Fund and cannot be altered by any one of the member countries more than 10 per cent without the consent of the Fund's directors. In order to maintain the exchange value of its currency, a member nation may borrow from the Fund whatever other currency it needs, the limit of its borrowing being set in proportion to its contribution. A great many complicated rules and regulations are part of the Monetary Fund's equipment, but in the main it is a device by which temporary scarcities of any currency in exchange markets are prevented from altering exchange rates. The Fund, however, is subject in the end to the same hazards that upset the international gold standard. If any nation's currency has a persistent tendency to fall, its borrowings from the Fund of a scarcer currency may reach its quota; in that event it would be in the same

plight as it was under the gold standard when it had to
stop further shipments of gold.

The International Bank for Reconstruction and Development was founded for a different purpose. Nations
may borrow from the Monetary Fund for temporary exchange needs, but many parts of the world also want long-
term loans to repair the ravages of war or to develop their
resources. One of the greatest requirements of the peoples
is an increase of production, which can be obtained only
with new investment in machinery and other capital
facilities. The International Bank is designed to encourage such investments on an international scale. Its capital is contributed by member nations in relation to their
economic capacity. The bank may lend (or guarantee
loans), up to a total limited by its capital, for purposes
which its directors regard as productive. This it may do
only when private capital is not available for the project
in question on reasonable terms. There have been, and
are likely to be, important opportunities for development
requiring long-term loans at relatively low rates which in
the end will probably be self-liquidating but which are
properly the task of governmental rather than of private
capital.

A third type of action also is needed to stimulate trade
and production on a world-wide scale. This is the removal
of all sorts of trade barriers which grew up during the
depression of the 1930's and were necessarily retained or
even strengthened for war purposes—tariffs, exchange
controls, export and import restrictions, bilateral barter
arrangements, and the like. The United States, through
the policy of its reciprocal trade agreements, strove with
some success to reduce or remove these barriers between
1934 and the outbreak of war in 1939. It had to suspend
the effort during the war, but it was successful in obtaining agreements on the part of other nations to abandon
obstructions when the emergency would be over. In the
fall of 1947 it obtained at Geneva a General Agreement

on Tariffs and Trade with twenty-two other important trading nations, embodying mutual concessions, and later it negotiated with a larger group of nations at Havana a still broader charter of a proposed International Trade Organization—at this writing still to be ratified. In general, these agreements permit the retention of some existing controls as long as they are required, but embody promises to abandon them when the emergency has passed. The charter of the ITO also contains provisions for international supervision of governmental agreements which set prices and purchases of commodities, like wheat and sugar, and regulation of private cartels, in the interest of the consumer as well as of the producer.

Scarce Dollars

Fitted out with a means of stabilizing exchange rates for the short run, an institution to encourage world-wide investment so as to increase production and income, and numerous agreements to remove governmental obstructions to international trade, the world, it would seem, could look forward to better times. Yet there is a deepseated maladjustment which none of these things, and not all three of them together, can greatly modify. Indeed, not one of them can operate as expected until this maladjustment is remedied.

One great obstruction to trade is a world-wide scarcity of dollars. This scarcity has existed for nearly a quarter century, though masked at times during that period. On the whole, it has been getting worse as time has passed.

Great Britain and the Continent need every year many things which they cannot obtain in sufficient quantity without buying them from the United States, Canada, and other countries in the Western Hemisphere where dollars are used for money. How can the necessary dollars be obtained? Only in four ways: (1) through purchases of British and European goods and services by the Western

Hemisphere; (2) by shipments of gold to the Western Hemisphere; (3) by earnings of European investments in the Western Hemisphere; or (4) by net capital movements from that Hemisphere, which may originate in sale of European securities or in loans or grants.

Not for many years have Britain and Europe sold as great a value of goods and services abroad as they have bought there. Before World War I they met this deficiency by the earnings of their foreign investments. This was true of their relationship to the United States in particular, for this country had been built up to a considerable extent by investments from across the Atlantic. But in that war the Western Allies had to buy huge quantities of munitions and supplies in the United States, and they had to pay for them partly by selling existing investments to Americans, partly by borrowing from the American people and government. After the war other nations for the first time in many years owed more money to Americans than Americans owed to them. This was true even if governmental war loans be disregarded.

There would have been, soon after that war, a critical shortage of dollars and an imminent collapse of the European economy if a temporizing development had not intervened. The United States, becoming prosperous, began to invest huge sums abroad, and particularly in Germany. While this increased foreigners' debt to us and so aggravated the long-term problem, the dollars going abroad temporarily provided foreigners with the purchasing power they needed to buy from us. Meanwhile, a high American protective tariff prevented Europe from coming closer to a balance of its current accounts by selling us more goods.

This adjustment worked passably as long as the stream of American foreign investment continued to flow. But the stream began suddenly to drop in 1928. In 1929 and for years thereafter the great depression in the United States cut down our orders of foreign goods as well. The resulting scarcity of dollars deepened world-wide depression and led to many of the governmental obstructions to trade which Americans deplored. Except in the case of Hitler Germany, these obstructions were largely measures of self-defense by foreign nations against the consequences of dollar scarcity. Not having enough dollars, they had to look for other means of buying what they needed; hence barter arangements, "clearing agreements," and the like were made. Quotas, exchange control, and the other limitations of foreign trade were measures to make sure of getting the most necessary imports. International barter arose when money was lacking, just as barter among individuals sprang up in the United States when, during the depression, money was not available.

World War II reduced British and European foreign investments to an absolute minimum. Lend-lease retarded this process as far as the United States was concerned, but even here the period of "cash-and-carry" before lend-lease went into effect compelled the British and French to sell large quantities of their American securities. In other parts of the world Britain not only sold capital holdings but incurred what amounted to a huge was debt. Even after World War I, in the years 1919 and 1920, Europe still had an income from foreign investment and "invisible" trade items equivalent to 4 billion dollars today; in 1946 and 1947 this same type of income was a minus quantity—the European deficit was 1.4 billion dollars.

Europe has been in the past, and still is, a great producer of wealth. She imports from the west only a minor percentage of what she consumes. Yet the imports she needs are vital. Without them her own production would fail. It is therefore necessary to find some way of obtaining all the dollars required. Europe is preparing to do this by increasing her production and selling more abroad. It appears to be the only way which can be depended upon in the long run. In the short run, the scarcity of dollars to sustain the European population and help build up more production must be supplied mainly by the United States.

As long as dollars are scarce, other nations must regulate exchange, imports, and exports. They must ration and allocate goods internally. The world cannot get back to "normal" freedom of exchange, normal foreign balances of payments, and normal flow of foreign investment until the scarcity of dollars is conquered. The time required to achieve this end may be long. There is an ugly chance that it may not be achieved at all.

Two footnotes must be added to this brief summary of the calamitous scarcity of dollars in the rest of the world. One is that rising prices in the United States greatly aggravate the difficulty. Foreign nations need still more dollars to buy here when our prices go up. It is true that the prices of what they have for sale rise as well, but in many cases these prices do not rise so much. The "terms of trade" for Great Britain become adverse in periods of high prices, favorable in periods of low prices.

The other footnote concerns an important aspect of the problem in which little improvement has been made, or is in prospect. Within the United States, a locality which does not have enough money to buy what it needs from a more prosperous part of the country, and cannot borrow it, is likely to lose population. People who are hungry and cold migrate to places where better paying jobs can be found, if there are such places. But migration across international boundaries has been so severely restricted that it amounts to a mere trickle. Freedom of trade and absence of currency troubles are consistent only with free movement of people. It is true that neither the British nor the French are eager to solve their problems by losing population. They may succeed without doing so. But in nations like Italy and Greece there seems

to be almost no hope for a solution unless the surplus
population can move elsewhere and unless the birth rate
falls.

Scarce Goods

Shortage of dollars is, in a sense, a regional problem
and may, with wise handling and good luck, turn out to
be serious for only a few years. Behind it there is a problem as old as the world, though one which has temporarily been obscured here and there by fortunate circumstances. It is the race between population and the means
of sustenance.

Strangely enough, through all the misery of destruction
and war, population kept right on increasing—in Europe,
Asia, and Africa as well as in continents which were never
bombed and on which no hostile army set foot. Nowhere
in the world are there enough food, clothing, houses, to
say nothing of less elementary necessities, to satisfy the demand for them. Even in the United States, incomparably
richer than other countries and now producing much
more than before the war, we know this to be true. Here
dollars are far from scarce; the abundance of dollars
merely reveals the scarcity of goods. The delusion of depression days—that there were "surpluses"—has been ex-
posed by common experience. The surpluses were merely
goods that people could not buy, not goods that were not
wanted.

The present scarcity of goods on a world scale has been
aggravated by special circumstances. In Europe the most
serious is the economic collapse of Germany, formerly one
of the most productive nations in the world. Without
the heavy goods and coal formerly turned out by the
Ruhr, the Rhineland, and Silesia, the rest of Europe is
impoverished. This is a simple statement of fact, however
much one may fear the military power of a strong Germany. In pre-war times the rest of Europe bought more

from Germany than it sold to Germans. Germany's surplus industrial production was a powerful help in maintaining Europe's living standards. England in turn bought more from Europe than it sold there. The triangular circuit of trade within Europe is broken; at one of its sides is the wide gap left by the disappearance of Germany as a surplus producer.

Another shortage of goods arises from new developments in Eastern Europe (including Eastern Germany). From here the Western manufacturing nations used to obtain much of their needed food and raw materials. They no longer can get as much and it is uncertain whether the trade can be fully resumed. Aside from political difficulties, and aside from war devastation, the breaking up of great estates and the substitution of small peasant proprietorship has reduced agricultural output of grains and has increased domestic consumption of wheat.

Finally, disturbances in Asia and other remote parts of the world have hampered production and export of goods on which industrial nations depend. Neither India nor China ever fed her own population sufficiently, nor was there ever much check to the growth of these populations except famine and disease, but war and political change have added to the difficulty. Malaya and Indonesia have also been in turmoil to achieve their freedom. Will these nations in the coming years produce enough goods for themselves, and will they be able to export increasing quantities? Perhaps the answer is yes, but it is a feeble perhaps.

A restoration of German production, a growth of trade between Eastern and Western Europe, and a period of productive advance in the Far East are three of the essential conditions for even beginning to supply the world's real need for goods. It is even uncertain, according to the Food and Agriculture Organization of the United Nations, whether enough food can be grown for many years to come.

Statements about deficiencies of the international income and disruptions of the system of exchange are not mere vague inferences. They are substantiated in a good deal of detail by applications of the new economic science of social accounting by the Research and Planning Division of the Economic Commission for Europe and other competent agencies. In spite of the fact that more of the relevant figures are lacking than in the case of national income studies, work on the international income is proceeding rapidly and has illuminated some of the more critical problems

The Industrial Revolution Has Just Begun

The industrial revolution—the substitution of machinery, the factory system, and mechanical power for handicrafts and self-sustaining farm households—used to be thought of as a change which took place in England more than a century ago, and somewhat later took hold in the United States and other Western countries. We are now, it is assumed, in a new era of machine and power production which changes the nature of all social problems. This view forgets the great masses of people on other continents and the great land areas in which the industrial revolution either has not begun or has found only a slight foothold.

Industrial development has been rapid in India, some Latin-American countries, and other backward regions, if one measures its growth against the tiny base with which it started. But measured against what it still has to achieve in order to make those regions as productive as the United States is today, it has barely begun its task. Unless the pace of development should be more swift than it was in North America, there is at least a century of work and growth ahead before the undeveloped regions can begin to catch up. The sale of machinery and other capital goods to these countries may help to solve Western Europe's problem of expanding exports.

The undeveloped countries, observing the wealth of the United States, are clamoring for rapid industrialization. Unfortunately, many of them do not understand that more is required than factories, tools, and the money to buy them. Industrial productivity is built upon education, moderately good standards of health and nutrition, an aggressive and objective attitude toward the problems of production, slowly and painfully learned "know how" on the part of managers, technicians, workers. There are cases of countries which have imported such things as an elaborate basic steel plant of the most modern design, without having supplied their farming population with elementary necessities like metal plows, pitchforks, wagons, or good roads, and without having any of the simple equipment required to produce these elementary necessities. To impose highly technical and complex industries from above by governmental fiat on a people unready for them by training, experience, or attitude, is like putting a frosting on an unbaked cake.

Even in the most advanced countries there is no sign that the industrial revolution is approaching its end. The pace of scientific discovery and technical application of science has not slackened, and a large part of the prevailing practice remains far behind the technical frontier.

It is in furthering the industrial revolution on a world-wide scale that our only hope of stable abundance lies. In this sense, Americans, who have lost the stimulus of their own frontier, can find enough foreign frontier regions to keep their surplus energies active for many years to come. The new frontier cannot be exploited on the basis of military conquest or by the extension of government from Washington, as was the old. Some of the regions to be developed are sparsely settled, as was North America, but most of them are already overfilled with people. The essentials for their progress are good local and world government, well-ordered and progressive economy in the already developed industrial centers, and education in its basic sense.

All this will require large staffs of people who already know much—some in general and others in detail—about how the world lives and makes its living. These people must keep on learning more. Far more is already known about the physical possibilities of high production than is yet put to use. But far less is known about the larger requirements of economic and political organization, and about the other aspects of group human behavior, than is needed to make good use of that technical knowledge, even by the most advanced nations. Economists are just beginning to bring their science abreast of the demands made on it. Here is one of the main critical areas of stress. The ability to produce more, and to produce it steadily, will determine whether civilization as we know it can go on to fulfill its promise, or will disintegrate. What happens will depend, first, on what we know, second, on how well we apply what we know. It would be well to follow closely the advance of economic science. Good management of the world economy, which is now urgent, can no longer be left to chance, ignorance, and prejudice.

Selected References

To FOLLOW CURRENT FIGURES OF NATIONAL INCOME, EMPLOYMENT, PRODUCTION, AND OTHER ECONOMIC STATISTICS, CONSULT:

The Economic Report of the President. Washington: U.S. Government Printing Office, January 1947, July 1947, January 1948, July 1948, and periodically thereafter.

The Survey of Current Business. U.S. Department of Commerce monthly.

The Federal Reserve Bulletin. Board of Governors of the Federal Reserve System monthly.

FOR MORE DETAILED MATERIAL ON NATIONAL INCOME:

National Income Supplement to the Survey of Current Business. July 1947.

Kuznets, Simon. *National Income and its Composition, 1919—1938.* New York: National Bureau of Economic Research, 1941.

————. *National Product Since 1869.* New York: National Bureau of Economic Research, 1946.

————. *National Income: A Summary of Findings.* New York: National Bureau of Economic Research, 1946.

Stigler, George J. *Trends in Output and Employment.* National Bureau of Economic Research, 1947.

FOR THE BUSINESS CYCLE:

Mitchell, Wesley C. *Business Cycles, The Problem and Its Setting.* New York: National Bureau of Economic Research, 1927.

Burns, A. F. and Mitchell, Wesley C. *Measuring Business Cycles.* New York: National Bureau of Economic Research, 1946.

FOR THE OPERATION OF THE BANKING SYSTEM:
The Federal Reserve System: Its Purposes and Functions. Washington: Board of Governors of the Federal Reserve System, 1947.

FOR THE KEYNES THEORIES:
Keynes, J. M. *The General Theory of Employment, Interest and Money*. New York: Harcourt, Brace, 1936.
Robinson, Joan. *Introduction to the Theory of Employment*. New York: Macmillan, 1937.
Terborgh, George. *The Bogey of Economic Maturity*. Chicago: The Machinery and Allied Products Institute, 1945.

FOR INDUSTRIAL CONCENTRATION:
Smaller War Plants Corporation (Report), Economic Concentration and World War II. Washington: U.S. Government Printing Office, 1946.
Berle, A. A. and Means, Gardiner C. *The Modern Corporation and Private Property*, New York: Macmillan, 1933.
Chamberlin, Edward. *The Theory of Monopolistic Competition*. Cambridge: Harvard University Press, 1946.
Knauth, Oswald. *Managerial Enterprise*. New York: Norton, 1948.

FOR THE INTERNATIONAL SITUATION:
Research and Planning Division, Economic Commission for Europe. *A Survey of the Economic Situation and Prospects of Europe*. United Nations, 1948.

History and Economics

Religion and the Rise of Capitalism *by R. H. Tawney.* The influences of religious thought on the social and economic structures of the world. (#MD163—50¢)

Mainsprings of Civilization *by Ellsworth Huntington.* A penetrating analysis of how climate, weather, geography, and heredity determine a nation's character and history. Diagrams, maps, tables, bibliography. (#MT248—75¢)

Three Essays on Population *by Thomas Malthus, Julian Huxley, Frederick Osborn.* Three timely and important essays that explore a critical world problem, the birth rate which continues to rise in a world already crowded with people. (#MD295—50¢)

The Rich and the Poor: The Economics of Rising Expectations *by Robert Theobald.* A challenging analysis of how the underdeveloped countries may raise their productivity and increase their share of the world's economic wealth. (#MD325—50¢)

Ethics in a Business Society *by Marquis W. Childs and Douglass Cater.* A challenging analysis of the nature of moral values in the modern world. (#MD107—50¢)

This Little Band of Prophets: The British Fabians *by Anne Fremantle.* A penetrating examination of the history and influence of the Fabian Society and the famous personalities who founded it, including G. B. Shaw, H. G. Wells, Bertrand Russell, and the Webbs. (#MT266—75¢)

Uses of the Past *by Herbert J. Muller.* The civilizations of the past, how they flourished, why they fell, and their meaning for the present crisis of civilization. (#MD112—50¢)

Islam in Modern History *by Wilfred Cantwell Smith.* A noted scholar discusses the impact of Mohammedanism on Middle Eastern political life today. (#MD268—50¢)

The Greek Experience *by C. M. Bowra.* An extraordinary study of Greek culture, its achievements and philosophy. 48 pages of photos. (#MP349—60¢)

American Skyline *by Christopher Tunnard and Henry Hope Reed.* A fascinating panorama of American civilization as shown in the growth of our cities and towns. Illustrated. (#MD175—50¢)

Democracy in America (abridged) *by Alexis de Tocqueville.* The classic critique of freedom and democracy in 19th century America by a brilliant Frenchman. (#MT362—75¢)

A Documentary History of the United States (revised, expanded) *edited by Richard D. Heffner.* Important documents that have shaped our history. (#MD78—50¢)

The Cycle of American Literature *by Robert E. Spiller.* A striking history of America's literary achievements from the beginning to the present. (#MP382—60¢)

American Essays (expanded) *edited by Charles B. Shaw.* A sample of American thought from the 18th century to the present by Emerson, Twain, and others. (#MP377—60¢)

The United States Political System and How it Works *by David Cushman Coyle.* A key to national, state and local politics, and theories of democracy. (#MD319—50¢)

The United Nations and How It Works (revised) *by David Cushman Coyle.* The only book of its kind in the paperbound field. Prepared in close cooperation with the Department of Public Information of the United Nations. (#MD318—50¢)

A Brief History of the United States *by Franklin Escher, Jr.* A convenient guide to United States history outlining events and trends that make our heritage. (Signet Key #KD367—50¢)

The American Presidency *by Clinton Rossiter.* A clear account of the history and evolution of the Presidency and the President's current responsibilities. (#MP361—60¢)

THE MENTOR PHILOSOPHERS

The entire range of Western speculative thinking from the Middle Ages to modern times is presented in this series of six volumes. Each book contains the basic writing of the leading philosophers of each age, with introductions and interpretive commentary by noted authorities.

"A very important and interesting series."
—*Gilbert Highet*

THE AGE OF BELIEF: The Medieval Philosophers *edited by Anne Fremantle* (#MD126—50¢)
"Highly commendable . . . provides an excellent beginning volume."
—*The Classical Bulletin*

THE AGE OF ADVENTURE: The Renaissance Philosophers *edited by Giorgio de Santillana* (#MD184—50¢)
"The most exciting and varied in the series."
—*New York Times*

THE AGE OF REASON: The 17th Century Philosophers *edited by Stuart Hampshire* (#MT367—75¢)
"His (Hampshire's) book is a most satisfactory addition to an excellent series."
—*Saturday Review*

THE AGE OF ENLIGHTENMENT: The 18th Century Philosophers *edited by Sir Isaiah Berlin* (#MD172—50¢)
"(Sir Isaiah) has one of the liveliest and most stimulating minds among contemporary philosophers."
—*N. Y. Herald Tribune*

THE AGE OF IDEOLOGY: The 19th Century Philosophers *edited by Henry D. Aiken* (#MD185—50¢)
". . . perhaps the most distinct intellectual contribution made in the series."
—*New York Times*

THE AGE OF ANALYSIS: 20th Century Philosophers *edited by Morton White* (#MT353—75¢)
"No other book remotely rivals this as the best available introduction to 20th century philosophy."
—*N. Y. Herald Tribune*

The Sciences

The Wellsprings of Life *by Isaac Asimov.* An introduction to the secrets of living beings as revealed by the chemistry of the cell. (#MD322—50¢)

The Individual and the Universe *by A. C. B. Lovell.* An outstanding expert tells how new techniques of radio-astronomy are modifying cosmic theories. (#MD330—50¢)

One Two Three . . . Infinity *by George Gamow.* Current facts and speculations of science presented clearly by a leading physicist. (#MD97—50¢)

The Creation of the Universe *by George Gamow.* A cogent and stimulating book about the origin and evolution of the universe, according to modern theory. (#MD214—50¢)

Science and the Modern World *by Alfred North Whitehead.* A penetrating study of the influence of three centuries of scientific thought on civilization. (#MD162—50¢)

On Understanding Science *by James B. Conant.* A noted atomic physicist explains the scope of science today against an historical view of its growth. (#MD68—50¢)

The Universe and Dr. Einstein *by Lincoln Barnett.* A clear analysis of time-space-motion concepts and the structure of atoms. Foreword by Albert Einstein. (#MD231—50¢)

The Sea Around Us *by Rachel L. Carson.* This National Book Award winner is an enthralling account of the ocean and its inhabitants. (#MD272—50¢)

Under the Sea Wind *by Rachel L. Carson.* Life among birds and fish on the shore, in the open sea, and on the sea bottom. (#MD128—50¢)

The Mentor Religious Classics

THE HOLY BIBLE IN BRIEF *edited by James Reeves*
The basic story of the Old and New Testaments told as one clear, continuous narrative. (#MD116—50¢)

THE PAPAL ENCYCLICALS *edited by Anne Fremantle*
The most important pronouncements of the Popes through the ages, in their historical context.
(#MT256—75¢)

THE LIVING TALMUD: The Wisdom of the Fathers and Its Classical Commentaries, selected and translated by Judah Goldin. (#MT286—75¢)

THE MEANING OF THE GLORIOUS KORAN: An Explanatory Translation by Mohammed Marmaduke Pickthal.
The complete sacred book of Mohammedanism, translated with reverence and scholarship. (#MQ375—95¢)

THE SONG OF GOD: Bhagavad-Gita
The Hindu epic translated by Swami Prabhavananda and Christopher Isherwood. (#MD103—50¢)

THE WAY OF LIFE: Tao Tê Ching *by Lao Tzu*
A new translation by R. B. Blakney of a masterpiece of ancient Chinese wisdom. ((#MD129—50¢)

THE SAYINGS OF CONFUCIUS
translated by James R. Ware
The wise teachings of the ancient Chinese sage in a new translation. (#MD151—50¢)

THE TEACHINGS OF THE COMPASSIONATE BUDDHA
edited by E. A. Burtt
The basic texts, early discourses, the Dhammapada, and later writings of Buddhism. (#MP380—60¢)

TO OUR READERS: We welcome your request for our free catalog of SIGNET and MENTOR books. If your dealer does not have the books you want, you may order them by mail, enclosing the list price plus 5¢ a copy to cover mailing. The New American Library of World Literature, Inc., P. O. Box 2310, Grand Central Station, New York 17, N. Y.